Postharvest

Dr Ron Wills is an Emeritus Professor in the School of Environmental and Life Sciences at the University of Newcastle, Australia. He has had 40 years experience in many aspects of postharvest horticulture and has published more than 280 research papers. He has held government, university and industry positions in Australia and New Zealand and has consulted for government and international agencies on postharvest development projects throughout Asia.

Dr Barry McGlasson is currently Adjunct Professor in the Centre for Plant and Food Science at the University of Western Sydney, Australia. He has had 50 years experience in the postharvest physiology and technology of horticulture crops, with extensive international experience in assistance projects conducted by the Australian Government and in liaison with overseas universities. He has also been a supply chain management consultant to growers, category managers, packers, exporters and major retailers.

Dr Doug Graham was a Chief Scientist in the CSIRO and formerly Head of the Food Research Laboratory and Acting Chief of the CSIRO Division of Food Science and Technology, Australia. He has had extensive experience in the biochemistry and physiology of horticultural crops and has advised on postharvest and research development in Asia and elsewhere.

Dr Daryl Joyce is Professor and Director of the Centre for Native Floriculture at the University of Queensland, Australia. He has worked for over 20 years in postharvest horticulture research, teaching and extension in Australia, USA and the UK. His research has focused on understanding the postharvest biology of native Australian cut flowers and foliage and on implementing postharvest technologies and systems to supply high-quality products to consumers.

Postharvest

An introduction to the physiology and handling of fruit, vegetables and ornamentals

5th edition

R.B.H. Wills • W.B. McGlasson
D. Graham • D.C. Joyce

Published in Australia, New Zealand, Papua New Guinea and Oceania by
University of New South Wales Press Ltd
University of New South Wales
Sydney NSW 2052 Australia
www.unswpress.com.au

And in the rest of the world by
CABI
CABI is a trading name of CAB International
CABI Head Office
Nosworthy Way
Wallingford
Oxfordshire OX10 8DE
UK

Tel: +44 (0)1491 832111
E-mail: cabi@cabi.org
Website: www.cabi.org

CABI North American Office
875 Massachusetts Avenue
7th Floor
Cambridge, MA 02139
USA

Tel: +1 617 395 4056
Fax: +1 617 354 6875
E-mail: cabi-nao@cabi.org

First published 1981
Second revised edition published 1982
Third revised edition 1989
Fourth revised edition 1998

National Library of Australia
Cataloguing-in-Publication entry

Wills, R. B. H. (Ronald Baden Howe).
Postharvest: an introduction to the physiology and handling of fruit, vegetables and ornamentals.

5th ed.
Bibliography.
Includes index.
ISBN 9780868409801
ISBN 0 86840 980 4.

1. Vegetables - Postharvest physiology. 2. Fruit - Postharvest physiology. 3. Plants, Ornamental - Postharvest physiology. I. McGlasson, Barry. II. Joyce, Daryl.

635.046
A catalogue record for this book is available from the British Library.
A catalogue record for this book is available from the Library of Congress, Washington, DC, USA.

ISBN 978 0 86840 980 1 (UNSW Press)
ISBN 978 1 84593 227 5 (CABI)
Printer Everbest, China

This book is printed on chlorine-free paper.

Contents

Preface

Since *Postharvest* was first published in 1981 the postharvest physiology and technology of fresh produce and ornamentals has become a mature profession. In advanced economies the postharvest professional provides the core knowledge essential for the efficient management of the supply chain extending from farmers to the ultimate consumer. Affluent economies now have the luxury of year-round supplies for many commodities, with considerable choice among items or categories of fresh produce and ornamentals. These products may be transported long distances to domestic or global markets. The emphasis in these economies is on meeting the needs of consumers to ensure repeat purchases. There is now a focus on food safety, quality assurance and perceived value for money. In the 1981 edition, one of the stated objectives of postharvest technology was to reduce the considerable losses that existed throughout the postharvest chain. While still important in 2006, outright losses of fresh produce in developed economies are now less important than loss of value if handling practices do not maintain quality as the produce moves along the distribution chain.

Despite the global economic advances, a large part of the world's population remains poor; for them, the purchase of fresh produce can be a major household cost. When rural populations are able to grow fresh produce to meet their own needs, with perhaps some surpluses to be sold to nearby communities, there is little need for special postharvest technologies because fruit and vegetables are often consumed on the day they are harvested. However, with the increasing drift of people in developing countries from rural areas to large cities, the challenge remains to expand food availability and reduce outright losses with simple, appropriate postharvest technologies.

This book begins with basic information about the structure of fruit, vegetables and ornamentals and how this influences their postharvest behaviour, then summarises key information about their composition, biochemistry, respiration and physiology. Managing produce temperature is the core technology for maintaining fresh quality. How this is achieved, and the influence of temperature on relative humidity and water loss, is discussed in depth. Fresh products are also susceptible to various pathogenic diseases that need to be identified and controlled. Technologies that are adjuncts to temperature management, including controlled and modified atmosphere storage, controlled ripening, packaging systems and transport are discussed in some detail. Also outlined in some detail are the principles underlying the development of modern, food safety–based quality assurance systems that also meet environmental concerns.

This text is an introductory primer for tertiary courses and for people working along the supply chain from the farm or orchard to retail stores. It should also be a useful resource for discerning consumers.

Acknowledgements

The authors gratefully acknowledge the following colleagues who assisted with the revision of specific sections of the book: Ms Clare Hamilton-Bate, National Program Manager, Freshcare Ltd; Mr Joseph Ekman, NSW Department of Primary Industries, Gosford (quality assurance); Dr Elena Lazar, NSW Department of Primary Industries, Gosford (postharvest pathology); Dr Donald Irving, The University of Queensland (postharvest biology and technology); Dr Rod Jordan, Dr Lindy Coates and Mr Tony Cooke, Department of Primary Industries and Fisheries, Queensland (postharvest pathology); and Dr Ian Ferguson, HortResearch, New Zealand (postharvest disorders). The assistance of Dr Jenny Ekman (NSW Department of Primary Industries, Gosford) in the preparation of the illustrations is gratefully acknowledged.

Abbreviations

ACC	1-aminocyclopropane-1-carboxylic acid
ADP	adenosine diphosphate
AOA	aminooxyacetic acid
ATP	adenosine triphosphate
AVG	L-2-amino-4-(2-aminoethoxy)-trans-3-butenoic acid or aminoethoxyvinylglycine
C	carbon (e.g. a 4-C molecule has 4 carbon atoms)
CA	controlled atmosphere
DNA	deoxyribonucleic acid
EFE	ethylene-forming enzyme
ERH	equilibrium relative humidity
EU	European Union
FAD	flavin adenine dinucleotide
$FADH_2$	reduced form of FAD
FAO	Food and Agriculture Organization of the United Nations
GA	gibberellin
GI	glycaemic index
Gy	Gray (unit of irradiation)
HACCP	hazard analysis and critical control points
J	joule (a measure of energy)
M	molar (concentration of solution related to the molecular weight of a compound)
mol.	mole (unit of weight related to the molecular weight of a compound)
MA	modified atmosphere

MAP	modified-atmosphere packaging
MB	methyl bromide
1-MCP	1-methylcyclopropene
MRL	maximum residue level
mRNA	messenger ribonucleic acid
N	unit of the amount of hydrogen or hydroxide ions in solution
NAD^+	nicotinamide adenine dinucleotide
NADH	reduced form of NAD^+
$NADP^+$	nicotinamide adenine dinucleotide phosphate
NADPH	reduced form of $NADP^+$
NIR	near infra-red
OPPP	oxidative pentose phosphate pathway
Pa	Pascal (unit of pressure)
PG	Polygalacturonase
pH	log value of hydrogen ion concentration
P_i	inorganic phosphate
ppb	parts per billion
ppm	parts per million
Q_{10}	temperature quotient (10°C)
RH	relative humidity
SAM	S-adenosyl-methionine
SOPP	sodium ortho-phenylphenate
STS	silver thiosulphate
TBZ	thiabendazole
TCA	tricarboxylic acid
UV	ultraviolet
VP	vapour pressure
VPD	vapour pressure deficit
v/v	volume per volume
WHO	World Health Organization
w/v	weight per volume

1 Introduction

Importance of fruit and vegetables as food

Fresh fruit and vegetables have been part of human diets since the dawn of history, although most societies have tended to value foods from animal sources more highly. Societies with largely or totally vegetarian diets, for religious or economic reasons, have had a greater dependence on fruit and vegetables. With the assistance of modern nutritional science, the profile of fruit and vegetables has risen considerably and health professionals, particularly in developed countries, are actively recommending increased consumption of fruit and vegetables and restricted consumption of animal foods.

The nutritional value of some fruits and vegetables was recognised in the early 17th century in England. One example is the ability of citrus fruit to cure scurvy, a disease widespread among naval personnel. While individual captains took advantage of this knowledge to maintain the health of their crews on long voyages, it was not until the late 18th century that the British Royal Navy issued a regular ration of lime juice to all sailors, leading to their nickname, 'limeys'.

Ascorbic acid (vitamin C) was not discovered as the ingredient responsible for preventing scurvy until the 1930s. It has since been shown to have a range of beneficial effects related to wound healing and as an antioxidant. There is now also considerable speculation about its possible action as an anti-viral and anti-cancer agent. Dietary sources of vitamin

C are essential, since humans lack the ability for its synthesis. All fruit and vegetables contain vitamin C; as a group, they are the major dietary source of the vitamin, supplying about 90 per cent of bodily requirements in virtually all countries.

Specific fruits and vegetables are also excellent sources of the provitamin A carotenoids, which are essential for maintenance of ocular health; and folic acid, which prevents certain anaemias. FAO and WHO have been actively promoting the use of home vegetable gardens for many years, as an inexpensive, readily available way to combat vitamin deficiency diseases in less developed regions.

The recent rise in nutritional importance of fruit and vegetables has been stimulated by a range of degenerative diseases prevalent in sedentary affluent societies, particularly in Western countries. Epidemiological evidence shows that communities who consume higher amounts of fruit and vegetables have lower incidences of such diseases. Fruit and vegetables is one of the five food groups used by nutritionists to promote a healthy diet; a common recommendation is for at least seven servings of fruit and vegetables to be consumed every day.

Concerns over obesity and coronary heart disease have led to the promotion of reduced levels of fat in the diet, while dietary fibre is considered to be beneficial in reducing or preventing a raft of medical conditions including colonic and rectal cancers, diabetes, diverticulitis, gallstones, haemorrhoids, hiatus hernia and varicose veins. Fruit and vegetables are generally low in fat and reasonably high in dietary fibre and are thus promoted as a substitute for animal-based foods and highly refined plant-based foods. There is now also considerable interest in determining the potential of fruit and vegetables to protect against various cancers, due to the antioxidant properties of a range of their constituents, including phenolic compounds and carotenoids such as ß-carotene and lycopene. However, clinical trials to date have not generated conclusive support for the anti-carcinogenic activity of any carotenoid.

The status of fresh fruit and vegetables has also benefited from an international trend towards fresh, natural foods; these are perceived to be superior to processed foods and to contain less chemical additives. This community perception has, however, placed additional pressure on the horticultural industry to retain its fresh, natural image by minimising the use of synthetic chemicals during production and postharvest handling.

Notwithstanding their nutritional status and their appeal as fresh and natural foods, the attraction of fruit and vegetables for many consumers is the sensory stimulation they impart. Fruit and vegetables provide variety in the diet through differences in colour, shape, taste, aroma and texture that distinguish them from the other major food groups of grains, meats and dairy products. The sensory appeal of fruit and vegetables is not confined to consumption but also has market value. Their colour and shape diversity is used to great effect by traders in arranging product displays to attract potential purchasers (Plate 1), and chefs have traditionally used fruit and vegetables to enhance the attractiveness of prepared dishes or table presentations. The use of parsley and similar herbs to adorn meat displays is widespread throughout the Western world, while fruit and vegetable carvings have become an art form in countries such as Thailand, where they are used as table ornaments.

While the nutritional composition of ornamental horticultural crops is inconsequential to consumers, flowers are increasingly being included in prepared mixed salads and therefore can make a limited contribution to the diet. However, the principal contribution of ornamentals is their provision of sensory pleasure and serenity, derived from the colour, shape and aroma of individual species. Apart from the more traditional home uses of garden plants and cut flowers, foliage and flowering plants are increasingly being used in the interiorscapes of commercial premises, including offices, hotels and restaurants. The importance of ornamentals in a society's cultural life should not be underestimated. Considerable commercial opportunities arise from their role in ceremonies such as weddings and funerals, in conveying messages on special occasions such as Mother's Day and Valentine's Day, as decorations in parades and rallies, and in art and creative pastimes, as reflected in the growth of ikebana schools. Ornamentals are given official national status too, since most countries have a flower as one of the symbols of state.

Horticultural production statistics

Fruit and vegetables

Worldwide production of fruit and vegetables has been increasing over many years, partly in response to population growth but also due to rising living standards in most countries and active encouragement by government health agencies of fruit and vegetable consumption. Table 1.1 shows that total world production of fruit and vegetables has doubled in the 23-year period of 1980–2003. The bulk of this increase was due to China, which has shown a 7-fold increase in production and in 2003 accounted for about one-third of world production. The second greatest producer is India, which now accounts for about 10 per cent of world production. While fruit and vegetable production in India doubled over the 23-year period, this was in proportion with the world increase. The increases in China and India are understandable given their large populations and rapid rates of economic growth. Most of the production growth from other countries has also been in developing countries, with only relatively small increases in traditional producers such as USA and Spain. France and Japan are no longer in the top 10 producers, with production declining from 21 million tonnes to 18 and 16 million tonnes, respectively, over the 23 years.

Table 1.1 Major national producers of fruit and vegetables

Country	Production (millions of tonnes)				Percentage of 2003 world production
	1980	1990	2000	2003	
China	67	150	388	483	37
India	56	76	119	128	10
USA	52	56	68	66	5
Brazil	23	36	43	42	3
Turkey	21	27	35	37	3
Italy	34	32	34	31	2
Spain	21	24	28	29	2
Iran	8	15	24	24	2
Mexico	12	16	23	24	2
Egypt	10	13	21	22	2

SOURCE FAO Statistical Yearbook 2004. FAO, Rome, 2005.

The international trade in fruit and vegetables grew rapidly until about 1990, to around US$50 billion, but little growth has occurred since that time. The recent lack of growth is probably due to the saturation of markets in developed countries, particularly within the European Union (EU), the major importing region for fruit and vegetables. Statistics on imports and exports of fruit and vegetables are distorted, because certain countries act as transit centres for imports, which are then re-exported to other countries in their region. An example is Belgium, which does not produce bananas but exports them to the value of $660 million – a value greater than that of Costa Rica (about $500 million), a major world producer. Table 1.2 lists the major countries involved in import and export. The 15 EU countries (as at 2000) were the major importers of fruit and vegetables, followed by USA and Japan. The EU and USA are also major exporters. Of the developing countries, China, Mexico and Turkey are important exporters. Many developing countries have targeted fruit and vegetables as a national export specialty. This strategy has been aided by the more relaxed trade barriers to fresh fruit and vegetables in developed countries compared to many other agricultural commodities.

Table 1.2 Major international traders in fruit and vegetables in 2000

Imports		Exports	
Country	**% of total world value**	**Country**	**% of total world value**
EU	25	USA	17
USA	20	EU	11
Japan	12	China	8
Canada	6	Mexico	7
		Turkey	4
		Canada	4

SOURCE E. Leguen de LaCroix, The Horticulture Sector in the European Union. European Commission, Directorate General for Agriculture, 2003.

Ornamentals

Reliable statistics on the production and trade in ornamentals are difficult to obtain. Nonetheless, it is well established that ornamentals comprise a very important sector of horticulture. Ornamental crops include cut flowers and foliage, flowering and foliage pot plants, bedding plants, and containerised shrubs and trees. These crops constitute part of the lifestyle horticulture industry, which also includes turf.

Worldwide consumption of floral crops was around US$25 billion per annum in 1990. In 1995 the total world markets for cut flowers and potted plants were suggested to be US$31 billion and US$19 billion, respectively. Total floriculture product exports in 2001 were considered to be US$7.3 billion, comprised of US$0.5 billion for bulbs, US$2.7 billion for plants, US$3.6 billion for cut flowers, and US$0.5 billion for cut foliage.

The world trade in ornamentals is dominated by Europe and most notably by the Netherlands. Rose and kalanchoe head the top 10 cut flower and pot plant lists, respectively, of lines supplied in 2005 through the Dutch auction system (Table 1.3). In 2001, exports of cut flowers were dominated by the Netherlands (US$3.7 billion) with the next major exporters being Colombia, Canada, Belgium and Italy, each at US$0.3–0.4 billion.

Table 1.3 Top 10 cut flowers and pot plants in the Dutch 'Bloemenveiling Aalsmeer' (VBA) auction system in 2005

Cut flower	Volume (millions of stems)	Pot plant	Volume (millions of units)
Rose	1979	Kalanchoe	31
Tulip	668	Hyacinth	15
Spray chrysanthemum	545	Phalaenopsis (orchid)	15
Gerbera	272	Dracaena (dragon tree)	14
Lily	141	Ficus	13
Freesia	126	Pot rose	13
Alstroemeria (Peruvian lily)	124	Pot chrysanthemum	12
Gypsophila (baby's breath)	73	Saintpaulia (African violet)	12
Hypericum	63	Daffodil	10
Carnation	57	Spathiphyllum (peace lily)	10

SOURCE Key Figures 2005, Bloemenveiling Aalsmeer: www.aalsmeer.com, accessed 23/6/2006.

Due to allied secondary industries (e.g. florists and plant hire businesses) that rely on fresh ornamentals of high quality, postharvest (postproduction) handling is a critical issue. Compared with fruit and vegetables, ornamentals have higher multiplier value in terms of their use by secondary industries. Relative added value multipliers of around 0.5, 0.4 and 8.0 have been proposed for the fruit, vegetable and ornamental industries, respectively.

Need for postharvest technology

Fruit, vegetables and ornamentals are ideally harvested at optimum eating or visual quality. However, since they are living biological systems they will deteriorate after harvest. The rate of deterioration varies greatly between individual products depending on their overall rate of metabolism, but for many it can be rapid. For simple marketing chains where produce is transferred from farm to end user within a short time period, the rate of postharvest deterioration is of little consequence. However, with the increasing remoteness of production areas from population centres in both developing and developed countries, the proliferation of large urban centres with complex marketing systems and the growth in international trading, the time from farm to market can be considerable. The deliberate storage of certain produce to capture a better return adds to this time delay between farm and end user, by extending the marketing period into times of shorter supply. Thus the modern marketing chain puts increasing demands on produce and creates the need for postharvest techniques that allow retention of quality over an increasingly longer period.

Extending the postharvest life of horticultural produce requires knowledge of all the factors that can lead to loss of quality or generation of unsaleable material. The field of study that adds to and uses this knowledge in order to develop affordable and effective technologies that minimise the rate of deterioration is known as postharvest. The increased attention afforded postharvest horticulture in recent years has come through the realisation that faulty handling practices after harvest can cause large losses of produce that required substantial inputs of labour, materials and capital to grow. Informed opinion now suggests that increased emphasis should be placed on conservation after harvest, rather than endeavouring to further boost crop production – as this would appear to offer a better return for the available resources of labour, energy and capital.

The actual causes of postharvest loss are many, but they can be classified into two main categories. The first of these is physical loss. Physical loss can arise from mechanical damage or pest and disease damage resulting in produce tissue being disrupted to a stage where it is not acceptable for presentation, fresh consumption or processing. Physical loss can also arise from evaporation of intercellular water, which leads to a direct loss in weight. The resulting economic loss is primarily due to the reduced mass of produce that remains available for marketing but can also be due to a whole batch of a commodity being rejected because of a small proportion of wasted items in the batch.

Loss of quality is the second cause of postharvest loss, and this can be due to physiological and compositional changes that alter the appearance, taste or texture and make produce less aesthetically desirable to end users.

The changes may arise from normal metabolism of produce (e.g. senescence) or abnormal events (e.g. chilling injury) arising from the postharvest environment. Economic loss is incurred because such produce will fetch a lower price. In many markets there is no demand for second class produce, even at a reduced price, which leads to a total economic loss even though the goods may still be edible.

In tropical regions, which include a large proportion of the developing countries, these losses can assume considerable economic and social importance. In developed regions, such as North America, Europe and Australia/New Zealand, postharvest deterioration of fresh produce is often just as serious, although often for different reasons. As the value of fresh produce may increase many times on its journey from the farm to the retailer, the economic consequences of deterioration at any point along the chain are serious. When farms are located near towns and cities, faulty handling practices are often less of a problem, because the produce is usually consumed before serious wastage can occur. Even in the tropical regions, production of some staple commodities is seasonal, and there is a need to store produce to meet requirements during the off-season. In industrialised countries and in countries that encompass a wide range of climatic regions, fresh fruit and vegetables are frequently grown at locations remote from the major centres of population. Thousands of tonnes of produce are now transported daily over long distances, both within countries and internationally. Fresh fruit and vegetables are important items of commerce, and there is a huge investment of resources in transport, storage and marketing facilities designed to maintain a continuous supply of these perishable commodities. Postharvest technology aims to protect that investment.

While the magnitude of losses of horticultural produce during postharvest and marketing operations are widely acknowledged to be considerable, few studies have accurately quantified these losses. Part of the difficulty in quantifying postharvest losses is identifying the actual steps in the postharvest chain where the loss was induced. It is not uncommon for a physical or metabolic stress to be imposed on produce but not visually evident until later in the marketing chain. For example, exposure to excess field heat after harvest can advance general senescence, but visible symptoms such as loss of green colour may not occur for days or weeks. Also, the visible cause of loss may not be the actual cause; for example, chilling injury of tomatoes is induced by prolonged storage at sub-optimal temperatures, but visual symptoms are usually mould growth on the damaged tissues and not the chilling injury itself.

The Inter-American Institute for Cooperation on Agriculture (IICA) has worked for many years to develop thorough assessments of postharvest systems that minimise the need for large scale quantitative measurements. Their work has resulted in the generation of a practical manual (the Commodity Systems Assessment Methodology tool, Postharvest Institute for Perishables, www.uidaho.edu/uipip/index.html) that can guide the systematic identification of postharvest problems within any horticultural situation. Application of the methodology, however, requires an interdisciplinary team approach, as knowledge of all the pre-production, production, harvest, postharvest and marketing operations that comprise any commodity handling system is required.

Postharvest technology

The ultimate role of postharvest technology is to devise methods by which deterioration of produce is restricted as much as possible during the period between harvest and end use, and to ensure that maximum market value for the produce is achieved. This requires a thorough understanding of the structure, composition, biochemistry and physiology of horticultural produce, as postharvest technologies will be mainly concerned with slowing down the rate of produce metabolism without inducing abnormal events. While there is a common underlying structure and metabolism, different types of produce vary in their response to specific postharvest situations. Appropriate postharvest technologies must be developed to cope with these differences. The variation in response can also be important between cultivars of the same produce and also often between different maturities, growing areas or seasons.

The principal weapon in the postharvest armoury relates to controlling the storage environment and handling conditions. Control over temperature is the most important environmental factor, as it affects the rate of postharvest deterioration from all causes. Fresh horticultural produce must be kept within a certain temperature range. The lower limit is the freezing point of plant tissues (about –2°C to 0°C) and the upper limit is the point at which plant tissues start to collapse (around 40°C). The effects of temperature are not uniform over the range. Moreover, there are time/temperature relationships, such that produce can withstand abnormally high or low temperatures for a short period. Thus, detailed knowledge of the responses of particular produce across the temperature range is essential in determining optimal storage temperature conditions. In general, the ideal postharvest temperature is just above the freezing point, where metabolism is slowest; but other factors, such as the onset of abnormal metabolism at reduced temperatures, can limit the use of low temperatures. The suppression of microbial growth at reduced temperatures is also a major consideration in many postharvest systems.

Other important environmental conditions are the concentration of certain gases and water vapour in the atmosphere around produce. Maintenance of a high relative humidity atmosphere is necessary to minimise water loss – a key quality factor, since wilted or shrivelled produce has a greatly reduced market value. The use of modified and controlled atmospheres, utilising elevated carbon dioxide and reduced oxygen levels from the normal atmosphere concentrations of around 0.033 per cent and 21 per cent, respectively, has been known for many years to beneficially affect produce metabolism; but the difficulties in adequately containing gas levels within the beneficial range has restricted the use of this technology to a few commodities. The recent development of plastic films with variable gas permeability and other atmosphere-control features has re-ignited interest in modified atmosphere storage. The presence of ethylene in the atmosphere has been of concern in the postharvest handling of ornamentals and unripe climacteric fruit for many years, because it promotes abscission, ripening and senescence, but its presence around non-climacteric fruit and vegetables is also important.

The major abnormal postharvest events are physiological disorders, arising from adverse postharvest and preharvest environmental conditions or mineral imbalances arising during growth; and microbial decay arising from a range of bacteria and moulds that can infect produce before and/ or after harvest. Apart from ensuring that produce is not exposed to the causative factors, control measures in the past have tended to focus on synthetic chemicals. However, with current consumer concerns, there is a trend towards the use of natural compounds or physical treatments. Postharvest insect infestation tends not to be a serious problem, except where the insect is subject to quarantine restrictions (e.g. fruit flies) – then it becomes a major technical and international and regional trade issue.

Apart from generating information on the effects of environmental conditions on particular produce, postharvest research must develop technology that is user-friendly and cost effective, to enable the scientific knowledge to have commercial value. Technical information can be used to either adapt existing technology, such as the refrigerated container, which created a mobile cool storage chamber; or to design new technology, such as vapour-heat treatments for insect disinfestation. Packaging design is one facet of technology development, since packages are required to protect their contents from physical damage and contamination, while at the same time meeting other marketing criteria.

Once loss of quality and wastage in the postharvest chain are under control, the next goal is responsiveness to market needs in terms of consumer expectations about quality, safety and presentation. This has forced marked changes in quality assessment, which was originally limited to grading operations in the packing house for size or weight, the removal of defects, and ensuring the correct labelling of containers. It has now become a total quality management (TQM) operation, to ensure that all market specifications are met and that the enterprise operates in the most efficient manner. Quality is thus linked with profitability, and the successful implementation of TQM requires a complete understanding of all factors that affect produce and the market environment in which produce is traded. Thus, quality management starts in the field and continues until produce reaches the end user. Staff training is now an integral part of quality management, as individuals or work teams are empowered to take responsibility for ensuring predetermined quality criteria are met.

Fresh fruit and vegetable produce that is cut or lightly processed, making the original commodity more convenient for consumers, has become an important and expanding presence on supermarket shelves. However, such processing invariably increases metabolism and renders produce more susceptible to microbial attack and adverse environmental conditions. It is often a challenge to develop the technology that retains quality in such products for the desired market period at a reasonable cost.

Applications of molecular biology are of increasing interest to researchers in overcoming specific postharvest problems. While a number of genetically modified (GM) products have been developed, such as tomatoes that remain firm over an extended period through inhibition of the polygalacturonase enzyme system, commercialisation has been almost non-existent as consumers remain resistant to accepting GM foods. There is likely to be less resistance to the development of GM ornamentals, since these are not consumed. Carnations with an extended shelf life through diminished production of ethylene have been developed, but again commercialisation is limited. There needs to be a marked demonstrable benefit to growers, traders and consumers before GM produce can have a reasonable chance of commercial success.

2 Structure and Composition

Structure

The fruits available commercially comprise various combinations of tissues that may include an expanded ovary, the seed and other plant parts such as the receptacle (apple, strawberry, cashew apple), bract and peduncle (pineapple). A dictionary definition of fruit is 'the edible product of a plant or tree, consisting of the seed and its envelope, especially the latter when juicy and pulpy'. However, a consumer definition of fruit would be 'plant products with aromatic flavours, which are either naturally sweet or normally sweetened before eating': they are essentially dessert foods. This consumer perception has resulted in various immature fruit such as zucchini, cucumber and beans, and even ripe fruit such as tomato, peppers (capsicum) and eggplant (aubergine) being considered as vegetables. These products, which have been referred to as fruit vegetables, may be consumed cooked or raw and are eaten either alone, in the form of a salad, or accompanying meat or fish dishes. However, as illustrated in Figure 2.1, common fruits are derived basically from an ovary and surrounding or associated tissues. Most of the exaggerated developments of certain parts of the fruit structure arose naturally but have been accentuated by breeding and selection to maximise the desirable features of each fruit and minimise the superfluous features. Naturally seedless cultivars of some fruits (e.g. banana, grape, navel orange) and others induced by breeding (e.g. watermelon) or management (e.g. persimmon) illustrate extreme development.

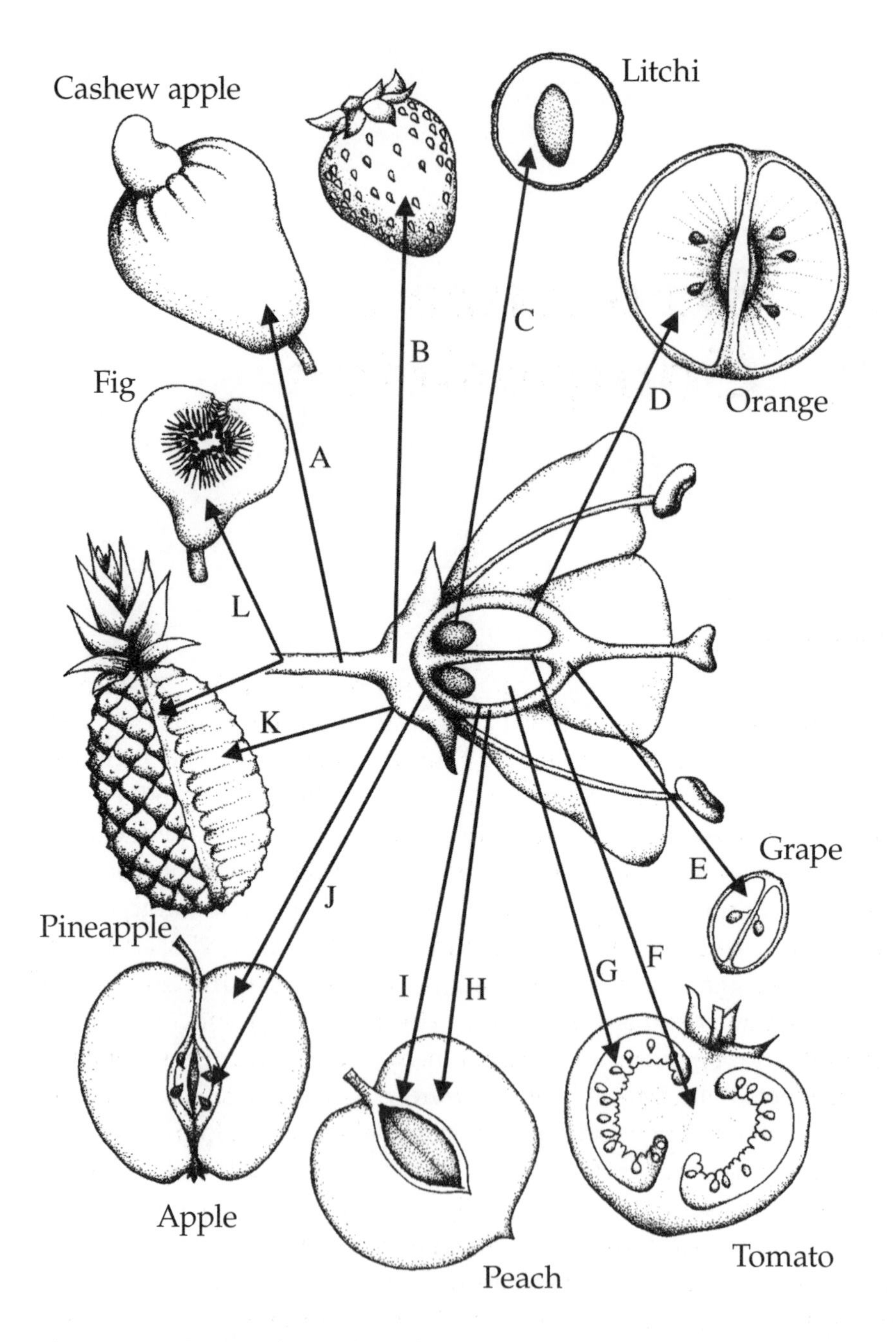

Figure 2.1 Derivation of some fruits from plant tissue. The letters indicate the tissues that comprise a significant portion of the fruit illustrated: (A) pedicel, (B) receptacle, (C) aril, (D) endodermal intralocular tissue, (E) pericarp, (F) septum, (G) placental intralocular tissue, (H) mesocarp, (I) endocarp, (J) carpels, (K) accessory tissue, (L) peduncle.

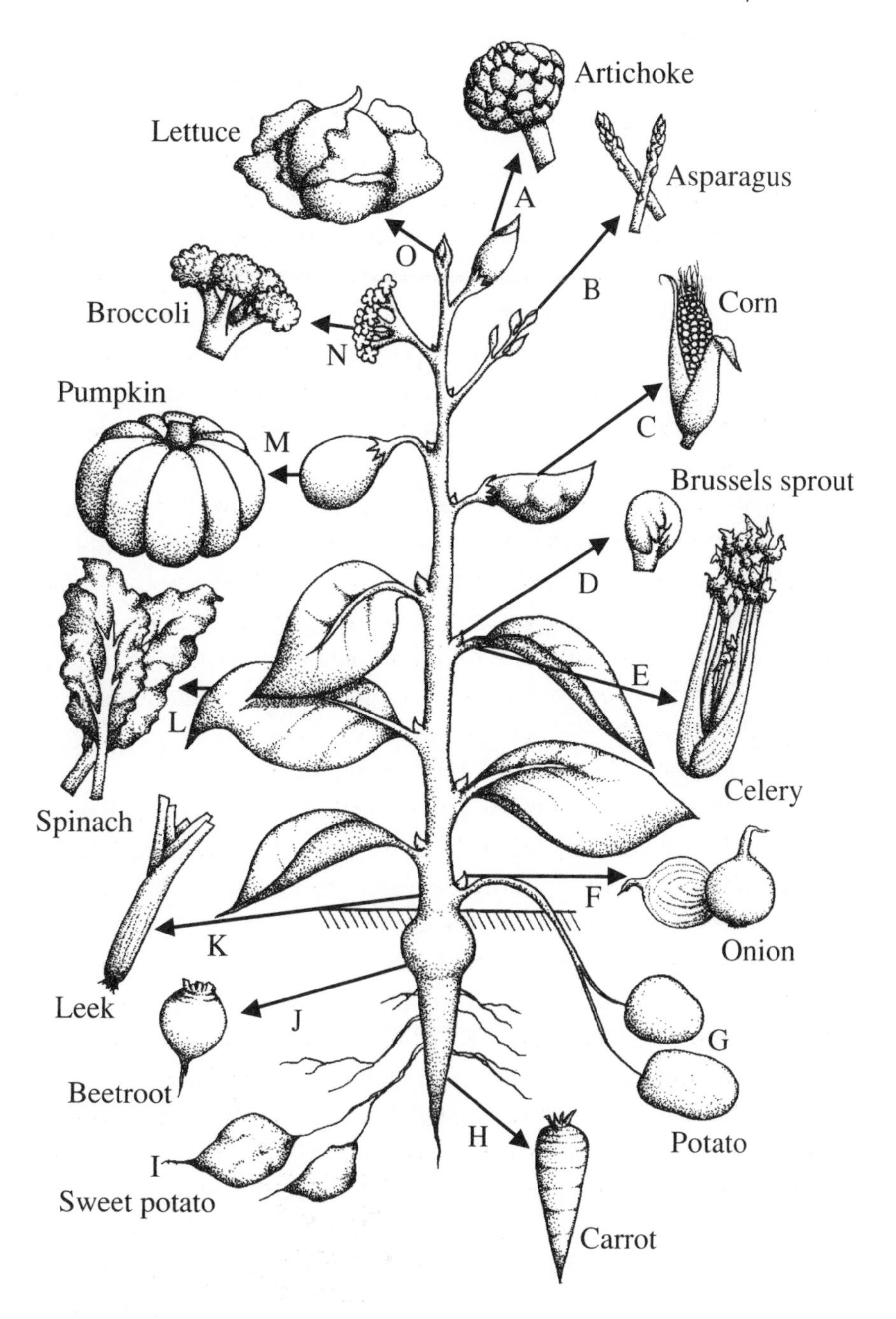

Figure 2.2 Derivation of some vegetables from plant tissue. The letters indicate the principal origins of representative vegetables as follows: (A) flower bud, (B) stem sprout, (C) seeds, (D) axillary bud, (E) petiole, (F) bulb (underground bud), (G) stem tuber, (H) swollen root, (I) swollen root tuber, (J) swollen hypocotyls, (K) swollen leaf base, (L) leaf blade, (M) fruit, (N) swollen inflorescence, (O) main bud.

The vegetables do not represent any specific botanical grouping, and exhibit a wide variety of plant structures. They can, however, be grouped into three main categories: seeds and pods; bulbs, roots and tubers; flowers, buds, stems and leaves. In many instances, the structure giving rise to the particular vegetable has been highly modified compared with that structure on the 'ideal' plant. The derivation of some vegetables is shown in Figure 2.2. The plant part that gives rise to the vegetable will be readily apparent when most vegetables are visually examined. Some are a little more difficult to categorise, especially the tuberous organs developed underground. The potato, for instance, is a modified stem structure, while other underground storage organs, such as the sweet potato, are simply swollen roots.

The structural origins of fruit and vegetables have a major bearing on the recommendations for their postharvest preservation. Generally, above-ground structures develop natural wax coatings as they mature, which reduce transpiration, whereas roots do not develop such coatings and therefore should be stored at high relative humidity (RH) to minimise water loss. Tuberous vegetables are equipped with a special capability to heal wounds caused by natural insect attack. This property is useful for minimising damage inflicted on tubers during harvesting.

Species of flowering plants used commercially as cut flowers have been selected for their visual appeal. At both the practical and botanical levels, cut flowers are variations of inflorescences. Although there is a wide range of variation in flower structure, the basic structure of an inflorescence is stem, including pedicels and peduncles, bract and flower. Figure 2.3 illustrates the range of variation in flower types. The range includes solitary to multiple inflorescences, where all inflorescences develop at about the same rate or where there is a gradation from mature to juvenile flowers as the inflorescence develops. It is important to recognise the variations in the structure of inflorescences, because they have a major bearing on postharvest handling strategies. Generally, inflorescences have a low carbohydrate reserve compared to most fruits, although it can be similar to that of many leafy vegetables. In many flower types, improved vase or storage life can be achieved by providing the cut flower with sugar through an absorbed solution. Because of their enormous surface area compared to their mass, cut flowers transpire at much higher rates than most fruit.

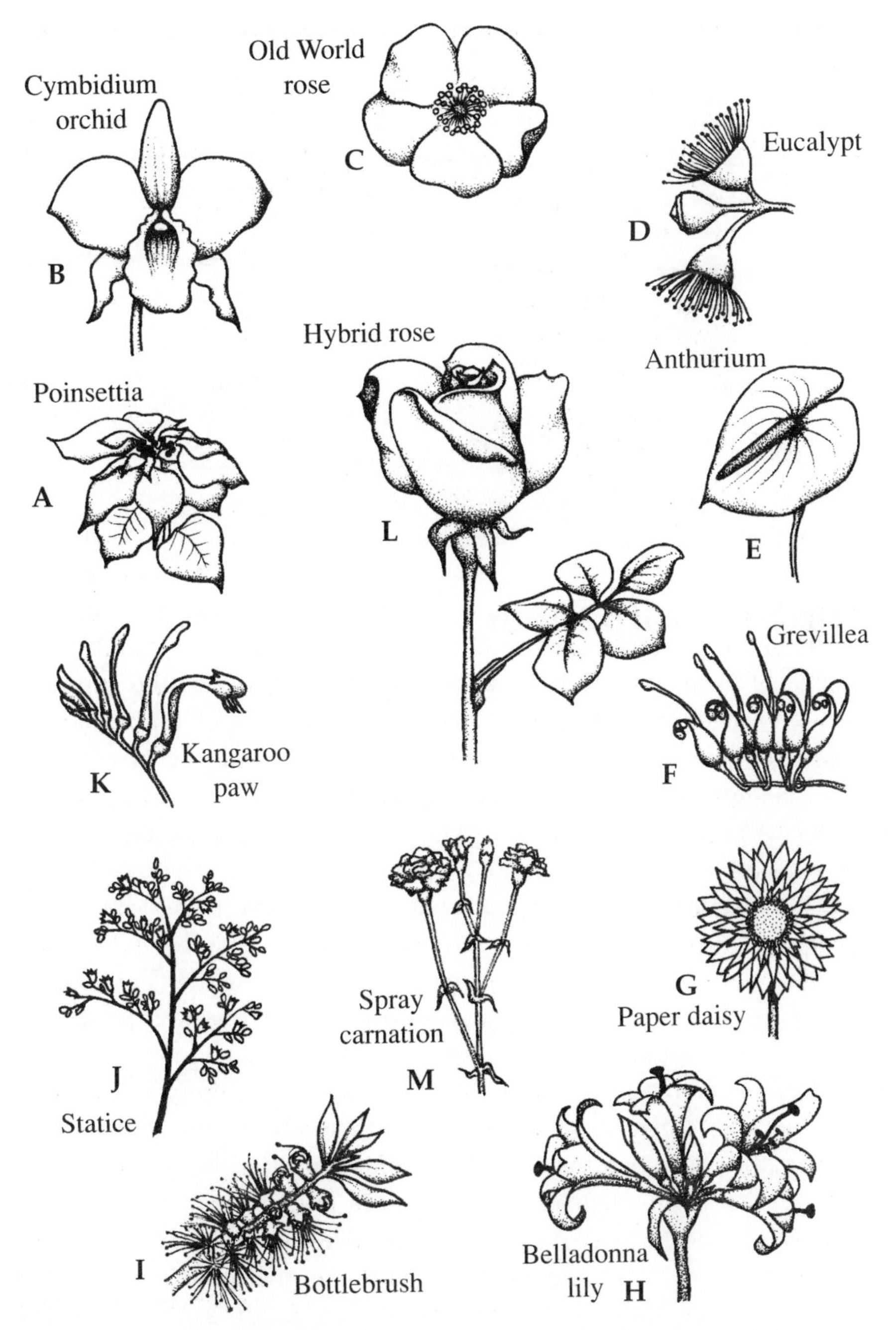

Figure 2.3 Examples of variations in the structure of flowers. (A) bract, (B) modifications and fusions, in which the labellum is a median modified petal and the column is comprised of fused stamens and pistils, (C) complete, single whorl of petals, (D) prominent feature (stamens), (E) spadix plus spathe, (F) raceme, (G) head, (H) umbel, (I) spike, (J) panicle, (K) syme, (L) solitary, (M) corymb.

Figure 2.4 Diagrammatic representation of a plant cell and its constituent organelles

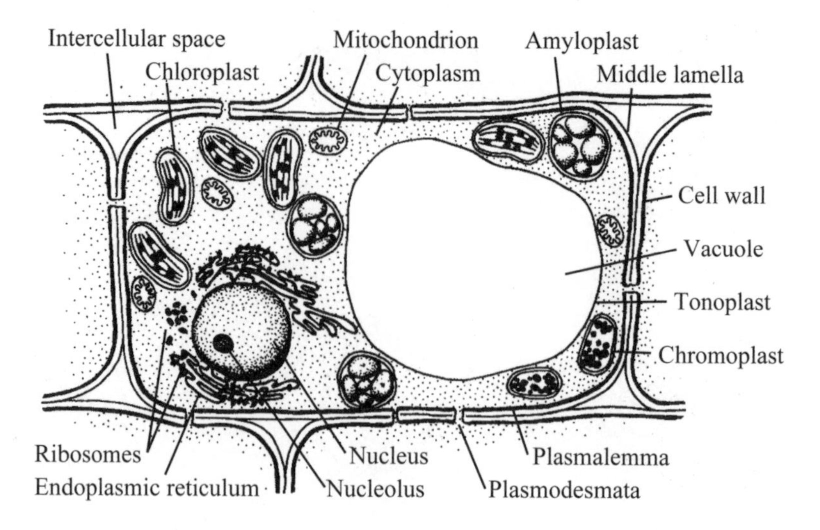

Cellular components

The cells of fruit and vegetables are typical plant cells, the principal components of which are shown in Figure 2.4. A brief outline of the essential features or functions of these components will be given; more detailed explanations can be found in specialised texts.

Plant cells are bounded by a mostly rigid cell wall composed of cellulose fibres, and other polymers such as pectic substances, hemicelluloses, lignins and proteins. A layer of pectic substances forms the middle lamella and acts to bind adjacent cells together. Adjacent cells often have small communication channels, called plasmodesmata, linking their cytoplasmic masses. The cell wall is permeable to water and solutes. Important functions of the wall are to support the cell membrane, the plasmalemma, against the hydrostatic pressures of the cell contents, which would otherwise burst the membrane; and to give structural support to the cell and the plant tissues.

Within the plasmalemma, the cell contents comprise the cytoplasm and usually one or more vacuoles. The latter are fluid reservoirs containing various solutes, such as sugars, amino and organic acids, and salts, and are surrounded by a semipermeable membrane, the tonoplast. Together with the semi-permeable plasmalemma, the tonoplast is responsible for maintaining the hydrostatic pressure of the cell, allowing the passage of water, but selectively restricting the movement of solutes or macromolecules, such as proteins and nucleic acids. The resulting turgidity of the cell is responsible for the crispness in fruit and vegetables.

The cytoplasm comprises a fluid matrix of proteins and other macromolecules and various solutes. Important processes that occur in this fluid part of the cytoplasm include the breakdown of storage reserves of carbohydrate by glycolysis (see Chapter 3) and protein synthesis. The cytoplasm also contains several important organelles, which are membrane-bound bodies with specialised functions as follows:

1. The nucleus, the largest organelle, is the control centre of the cell and contains the genetic information in the form of DNA (deoxyribonucleic acid). It is bounded by a porous membrane that has distinct holes, which can be observed through an electron microscope. These holes permit the movement of mRNA (messenger ribonucleic acid), the transcription product of the genetic code of DNA, into the cytoplasm where mRNA is translated into proteins on the ribosomes of the protein synthesising system (see below).
2. Mitochondria contain the respiratory enzymes of the tricarboxylic acid (TCA) cycle (see Chapter 3) and the respiratory electron transport system, which synthesise adenosine triphosphate (ATP). Mitochondria utilise the products of glycolysis for energy production. Thus they form the energy powerhouse of the cell.
3. Chloroplasts, found in the cells of green parts of a plant, are the photosynthetic apparatus of the cell. They contain the green pigment chlorophyll and the photochemical apparatus for converting solar (light) energy into chemical energy. As well, they have the enzymes necessary for fixing atmospheric carbon dioxide to synthesise sugars and other carbon compounds.
4. Chromoplasts develop mainly from mature chloroplasts when the chlorophyll is degraded. They contain carotenoids, which are the yellow-red pigments in many fruits.
5. Amyloplasts are the sites of starch grain development, although starch grains are also found in chloroplasts. Collectively chloroplasts, chromoplasts and amyloplasts are known as plastids.
6. The Golgi complex is a series of plate-like vesicles that bud off smaller vesicles that are probably of importance in cell wall synthesis and secretion of enzymes from the cell.
7. The endoplasmic reticulum is a network of tubules within the cytoplasm. Some evidence suggests they may act as a transport system in the cytoplasm. The endoplasmic reticulum often has ribosomes attached to it, which are the sites of protein synthesis. Other ribosomes are found free in the cytoplasm.

Chemical composition and nutritional value of fruit and vegetables

As a result of the importance of fruit and vegetables to health, there is now considerable information on their composition, particularly in terms of constituents with nutritional value. This data can be readily accessed through the many national and regional tables of food composition available on paper or electronically. However, care should be taken in applying such data due to the considerable differences in composition between cultivars and the effects of maturity, season, locality and storage. All of the following values are expressed on a fresh weight basis.

Water

Most fruits and vegetables contain more than 80 g of water per 100 g, with some tissues, such as cucumber, lettuce and melons, containing about 95 g/100 g. The starch tubers and seeds, for example yam, cassava and corn, contain less water, but even they usually comprise more than 50 g/100 g. Quite large variations in water content can occur within a species, since the water content of individual cells varies considerably. The actual water content depends on the availability of water to the tissue at the time of harvest, which varies during the day in response to fluctuations in temperature and relative humidity. For most produce, it is desirable to harvest when the maximum possible water content is present, as this results in a crisp texture. Hence the time of harvest can be an important consideration, particularly with leafy vegetables, which exhibit large and rapid variations in water content in response to changes in their environment.

Carbohydrates

Carbohydrates are generally the most abundant group of constituents after water. They can be present across a wide molecular weight range, from simple sugars to complex polymers that may comprise many hundreds of sugar monomeric units. Carbohydrates can account for 2–40 g/100 g of produce tissue, with low levels being found in some cucurbits, for example cucumber, and high levels in vegetables that accumulate starch, for example cassava.

The main sugars present in fruit and vegetables are sucrose, glucose and fructose, with the predominant sugar varying in different produce. Glucose and fructose occur in all produce and are often present at a similar level, while sucrose is only present in about two-thirds of produce. Produce with the highest sugar levels (Table 2.1) is mainly tropical and sub-tropical fruit, with grape the only temperate fruit listed and no vegetables listed. Beetroot contains the highest sugar content among the vegetables at about 8 g/100 g, with sucrose the only sugar present. Much of the sensory appeal of fruit is due to the sweet taste produced by sugars, which is considered to be one of the universal innate human taste preferences.

Table 2.1 Fruit and vegetables with the highest sugar levels (g/ 100 g)

Produce	Total sugars	Glucose	Fructose	Sucrose
Banana	17	4	4	10
Jackfruit	16	4	4	8
Litchi (lychee)	16	8	8	1
Persimmon	16	8	8	0
Rambutan	16	3	3	10
Grape	15	8	8	0
Custard apple	15	5	6	4
Pomegranate	14	8	6	0
Carambola	12	1	3	8
Mango	12	1	3	8

NOTE Some differences arise between values given for total sugars and the total of individual sugars, due to rounding of data given in R.B.H. Wills (1987) Composition of Australian fresh fruit and vegetables. *Food Technology Australia* 39, pp. 523–6.

Humans can digest and utilise sugars and starch as energy sources, hence vegetables with a high starch content are important contributors to the daily energy requirement of people in many societies. Produce such as cassava and yam commonly contain >20 g/100 g as starch, with other starchy produce containing >10 g/100 g starch. While an over-dependence on starchy vegetables is undesirable, as they cannot supply enough of certain essential nutrients, consumers in developed countries are being encouraged to eat more starch, or complex carbohydrate as it is now called, although in these countries cereals rather than fruit and vegetables are the major source of dietary starch.

Concern over the rise of diabetes in many societies has focused attention on the amount and type of carbohydrates in the diet and how they affect blood glucose levels. A glycaemic index (GI) of foods is now well established, and on this scale a pure glucose solution is rated as 100. The GI of fruit and vegetables varies widely and ranges from 22 (cherries) to 97 (parsnip). An interesting fact is that white bread, a common source of starch in Western countries, has a GI of 70 whereas starchy vegetables such as potato and sweet potato have a GI of 55–60.

A substantial proportion of carbohydrate is present as dietary fibre, which is not digested in the human upper intestinal system but is either metabolised in the lower intestines or passes from the body in the faeces. Cellulose, pectic substances and hemicelluloses are the main carbohydrate polymers that constitute fibre. Lignin, a complex polymer of aromatic compounds linked by propyl units, is also a major component of dietary fibre. Dietary fibre is not digested, as humans are not capable of secreting the enzymes necessary to break down the polymers to basic monomeric units that could be absorbed by the intestinal tract. Starch and cellulose have the same chemical composition, as they are synthesised from D-glucose units, but the bonding between the monomers differs. Starch comprises α-1,4 linkages, which are hydrolysed by a range of amylase enzymes secreted by humans; cellulose is formed with ß-1,4 linkages, however, the cellulase enzymes are not produced by humans. Similarly the enzymes necessary to hydrolyse the pectins and hemicelluloses to smaller units are lacking in humans. Dietary fibre was once considered to be an unnecessary component in the diet, although it was known to relieve constipation. But increased consumption of dietary fibre is now actively promoted by health agencies.

Protein

Fresh fruit and vegetables generally contain about 1 g/100 g protein in fruit and about 2 g/100 g in most vegetables, with the most abundant protein sources being the Brassica vegetables, which contain 3–5 g/100 g, and the legumes at about 5 g/100 g protein. The protein is mostly functional, for example in enzymes, rather than a storage pool as in grains and nuts. The relatively low level of protein means that fresh fruit and vegetables are not an important protein source in the diet.

Lipids

Lipids comprise less than 1 g/100 g of most fruit and vegetables and are associated with protective cuticle layers on the surface of the produce and with cell membranes. The avocado and olive (used as a fresh vegetable) are exceptions, having respectively about 20 and 15 g/100 g lipid present as oil droplets in the cells. The generally low lipid content of fruit and

Table 2.2 Levels of vitamin C, vitamin A and folic acid in some fruits and vegetables

Commodity	Vitamin C (mg/100 g)	Commodity	Vitamin A ß-carotene equivalent (mg/100 g)	Commodity	Folic acid (μ g/ 100 g)
Black currant, guava	200	Carrot	10.0	Spinach	80
Chilli	150	Sweet potato (red)	6.8	Broccoli	50
Broccoli, Brussels sprout	100	Parsley	4.4	Brussels sprout	30
Papaya	80	Spinach	2.3	Cabbage, lettuce	20
Kiwifruit	70	Mango	2.4	Banana	10
Citrus, strawberry	40	Red chilli	1.8	Most fruits	<5
Cabbage, lettuce	35	Tomato	0.3		
Banana, potato, tomato	20	Banana	0.1		
Apple, peach	10				

vegetables is seen as a positive factor in combating the rise of heart disease, and increased consumption of these foods is extensively promoted by health authorities. Even produce, such as avocado, with a relatively high lipid content contains mostly monounsaturated fatty acids. In recent years, monounsaturated fatty acids have been upgraded in nutritional importance because the 'Mediterranean diet', which is associated with a high consumption of olive oil, and hence of monounsaturated fatty acids, is now considered protective against heart disease.

Organic acids

Most fruit and vegetables contain organic acids at levels in excess of that required for the operation of the TCA cycle and other metabolic pathways. The excess is generally stored in the vacuoles away from other cellular components. Lemon, lime, passionfruit and black currant often contain over 3 g/100 g of organic acids. The dominant acids in produce are usually citric and malic acids. Other dominant organic acids in certain commodities include tartaric acid in grapes, oxalic acid in spinach and isocitric acid in blackberries. Apart from their biochemical importance, organic acids contribute greatly to taste, particularly of fruit with a balance of sugar and acid.

Table 2.3 Relative concentration of ten vitamins and minerals in fruit and vegetables and the relative contribution of vitamins and minerals these commodities make to the US diet

Nutrient concentration		Contribution of nutrients to diet	
Crop	**Rank**	**Crop**	**Rank**
Broccoli	1	Tomato	1
Spinach	2	Orange	2
Brussels sprout	3	Potato	3
Lima bean	4	Lettuce	4
Pea	5	Sweet corn	5
Asparagus	6	Banana	6
Artichoke	7	Carrot	7
Cauliflower	8	Cabbage	8
Sweet potato	9	Onion	9
Carrot	10	Sweet potato	10
Sweet corn	12	Pea	15
Potato	14	Spinach	18
Cabbage	15	Broccoli	21
Tomato	16	Lima bean	23
Banana	18	Asparagus	25
Lettuce	26	Cauliflower	30
Onion	31	Brussels sprout	34
Orange	33	Artichoke	36

SOURCE Adapted from C.M. Rick (1978) The Tomato. *Scientific American* 239(2), pp. 66–76.

Vitamins and minerals

Vitamin C (ascorbic acid) is only a minor constituent of fruit and vegetables but is of major importance in human nutrition for the prevention of the disease scurvy and for its possible protective antioxidant properties. Virtually all human dietary vitamin C (approximately 90%) is obtained from fruit and vegetables. The daily requirement for vitamin C is about 50 mg, and many commodities contain this amount of vitamin C in less than 100 g of tissue.

Fruit and vegetables may also be important nutritional sources of vitamin A and folic acid, commonly supplying about 40 per cent of daily requirements. Vitamin A is required by the body to maintain the structure of the eye; a prolonged deficiency of vitamin A can eventually lead to blindness. The active vitamin A compound, retinol, is not present in produce, but some carotenoids such as ß-carotene can be converted to retinol by humans. Only about 10 per cent of the carotenoids known to be in fruit and vegetables are precursors of vitamin A. All other carotenoids, such as lycopene, the main pigment in tomato, have no vitamin A activity but may be important as antioxidants.

Folic acid is involved in RNA synthesis, and a deficiency will result in anaemia. Folic acid deficiency during early pregnancy has been associated with foetal spina bifida, and various countries have moved to fortify certain foods with folic acid to minimise the risk. Green leafy vegetables are good sources of folic acid, with the intensity of green colour acting as a good guide to the folic acid content. Table 2.2 (page 23) shows the range of levels of vitamins C and A and folic acid in some fruit and vegetables. Maintenance of these vitamins during handling and storage should be a major concern, particularly when the produce will be consumed by people on marginally sufficient diets.

The major mineral in fruit and vegetables is potassium, which is present at more than 200 mg/100 g in most produce. The highest levels are in green leafy vegetables, with parsley containing about 1200 mg/100 g, but about 20 vegetables contain 400–600 mg/100 g. Health authorities in many countries are urging increased consumption of potassium to counter the effects of sodium in the diet, and fruit and vegetables are the richest natural food source of potassium.

Many other vitamins and essential minerals are present in fruit and vegetables, but their contribution to total dietary requirements is generally of minor importance. Iron and calcium may be present at nutritionally significant levels, although often in a form that is unavailable for absorption by humans; for example, most of the calcium in spinach is present as calcium oxalate, which is only poorly absorbed.

The nutritional value of various fruits and vegetables depends not only on the concentration of nutrients in the produce but also on the amount of such produce consumed in the diet. An attempt to equate these factors and show the relative concentration of ten major vitamins and minerals in some fruits and vegetables and their importance in the typical US diet in the 1970s is shown in Table 2.3. Tomatoes and oranges contain a relatively low concentration of nutrients but make the major contribution of all produce to the US diet because of the large per capita consumption.

Volatiles

All fruit and vegetables produce a range of small molecular-weight compounds (molecular weight less than 250) that possess some volatility at ambient temperatures. These compounds are not important quantitatively (normally less than 10 mg/100 g are present), but they are important in producing the characteristic flavour and aroma of fruit, and to a lesser extent of vegetables. Aroma constituents of many ornamentals have found application as perfumes for human use or as background odours in manufactured products. Most fruits and vegetables and many ornamentals each contain more than 100 different volatiles, and the number of compounds identified in particular produce is continually increasing as the sensitivity of the analytical techniques for their identification improves. The compounds are mainly esters, alcohols, acids and carbonyl compounds (aldehydes and ketones). Many of these compounds, such as ethanol, are common to all fruit and vegetables, while others are specific to an individual or species; esters, for example, are common constituents of most ripe fruits, while sulphur-containing volatiles are present in Brassica vegetables and tomatoes.

Definitive studies correlating consumer recognition of the produce with the volatile profile emanating from the produce have shown that only a small number of compounds are responsible for consumer recognition of that commodity. In most fruit and vegetables the characteristic aroma is due to the presence of one or two compounds. Table 2.4 gives the key compounds claimed to be responsible for the characteristic aromas of some fruit and vegetables. Practically all the compounds mentioned in Table 2.4 are minor components of the aroma fraction. The olfactory senses are thus extremely sensitive. The threshold concentration – the minimum concentration – at which the odour of ethyl 2-methylbutyrate, the main characteristic odour of apple, can be detected organoleptically was found to be 0.001 μL/L; that is, an apple weighing 100 g is recognised if 0.01 μg of ethyl 2-methylbutyrate is present. For the characteristic odour to be desirable, it must also be in the correct concentration. At different stages of maturation, different compounds become the dominant component of flavour. Thus a blindfolded subject would be able to detect the stage of development of a particular commodity by sniffing the aroma. Some examples of typical aroma compounds are given in Table 2.4.

Table 2.4 Distinctive components of the aroma of some fruits and vegetables

Product		Compound/s
Apple	ripe	Ethyl 2-methylbutyrate
	green	Hexanal, 2-hexenal
Banana	green	2-hexenal
	ripe	Eugenol
	overripe	Isopentanol
Grapefruit		Nootakatone
Lemon		Citral
Orange		Valencene
Raspberry		l-(π-hydroxyphenyl)-3-butanone
Cucumber		2,6-nonadienal
Cabbage	raw	Allyl isothiocyanate
	cooked	Dimethyl disulphide
Mushroom		l-octen-3-ol, lenthionine
Potato		2-methoxy-3-ethyl pyrazine, 2,5-dimethyl pyrazine
Radish		4-methylthio-*trans*-3-butenyl isothiocyanate

SOURCE Adapted from D.K. Salunkhe and J.Y. Do (1977) Biogenesis of aroma constituents of fruits and vegetables. *CRC Critical Reviews of Food Science and Nutrition* 8, pp. 161–90.

3 Physiology and Biochemistry

An important yet basic fact regarding the postharvest handling of horticultural produce is that fruit, vegetables and ornamentals are 'living' structures. We can readily accept that produce is a living, biological entity when it is attached to the growing parent plant in its agricultural environment. But even after harvest, the produce is still living: it continues to perform most of the metabolic reactions and maintain the physiological systems that were present when it was attached to the plant.

An important feature of plants, and therefore of fruit, vegetables and ornamentals, is that they respire by taking up oxygen and giving off carbon dioxide and heat. They also transpire (lose water). While attached to the plant, the losses due to respiration and transpiration are replaced from the flow of sap, which contains water, photosynthates (principally sucrose and amino acids) and minerals. Respiration and transpiration continue after harvest, and since the produce is now removed from its normal source of water, photosynthates and minerals, the produce depends entirely on its own food reserves and moisture content. Therefore, losses of respirable substrates and moisture are not made up and deterioration has commenced. In other words, harvested fruit and vegetables and ornamentals are perishable.

This chapter will consider the postharvest behaviour of horticultural commodities with particular reference to the physiological and biochemical changes that occur in ripening fruits. For this discussion, some understanding of the physiological development of fruit, vegetables and ornamentals is necessary.

Physiological development

The life of fruit and vegetables can be conveniently divided into three major physiological stages following initiation or germination. These are growth, maturation, and senescence (Figure 3.1). However, clear distinction between the various stages is not easily made. Growth involves cell division and subsequent cell enlargement, which accounts for the final size of the produce. Maturation usually commences before growth ceases and includes different activities in different commodities. Growth and maturation are often collectively referred to as the development phase. Senescence is defined as the period when anabolic (synthetic) biochemical processes give way to catabolic (degradative) processes, leading to ageing and finally death of the tissue. Ripening, a term reserved for fruit, is generally considered to begin during the later stages of maturation and to be the first stage of senescence. The change from growth to senescence is relatively easy to delineate. Often the maturation phase is described as the time between these two stages, without any clear definition on a biochemical or physiological basis.

Figure 3.1 Growth, respiration and ethylene production patterns of climacteric and non-climacteric plant organs

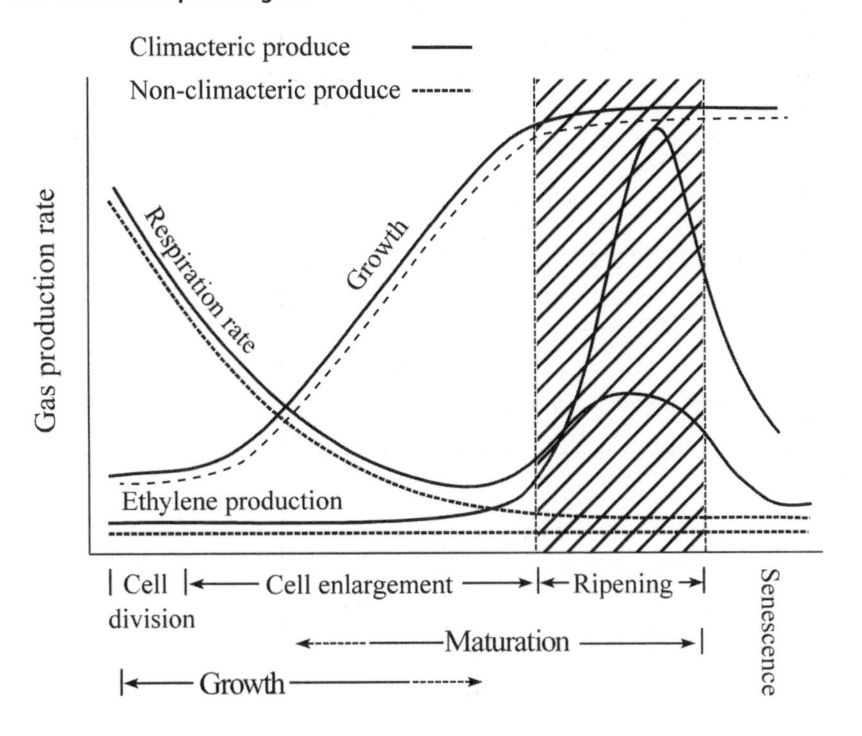

It is difficult to assign specific biochemical or physiological parameters to delineate the various stages, because the parameters for different commodities are not identical in their nature or timing. Figure 3.2 shows the major changes in certain biochemical and physiological parameters in the climacteric tomato as it ripens from mature green to ripe.

Figure 3.2 Physicochemical changes that occur during ripening of harvested tomatoes at 20°C

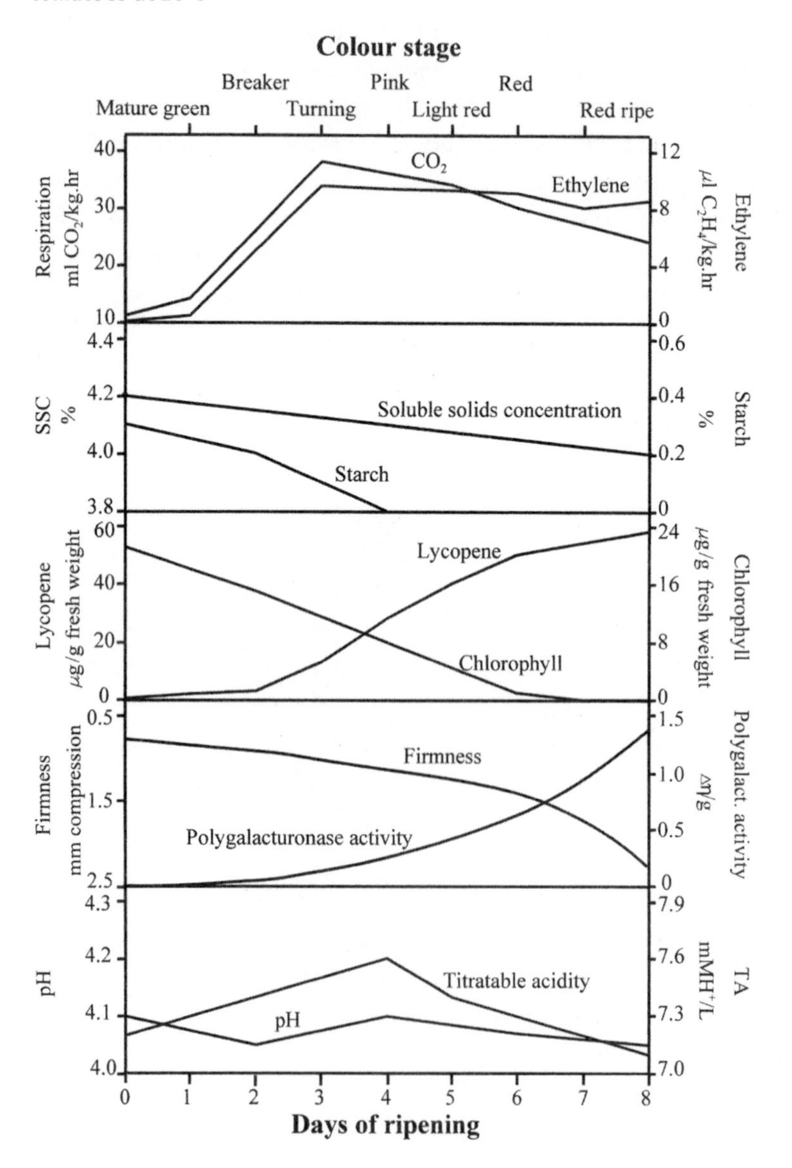

Development and maturation of fruit are completed only when it is attached to the plant, but ripening and senescence may proceed on or off the plant. Fruits are generally harvested either when mature or when ripe, although some fruits such as zucchini that are consumed as vegetables may be harvested even before maturation has commenced.

Similar terminology may be applied to vegetables and ornamentals, or to any determinant organ, except that the ripening stage does not occur. As a consequence it is more difficult to delineate the change from maturation to senescence in vegetables and ornamentals. Vegetables and ornamentals are harvested over a wide range of physiological ages; that is, from a time well before the commencement of maturation through to the commencement of senescence (see Figure 10.2).

Fruit ripening

Ripening fruit undergoes many physicochemical changes after harvest that determine the quality of the fruit eventually purchased by the consumer. Ripening is a dramatic event in the life of a fruit – it transforms a physiologically mature but inedible plant organ into a visually attractive olfactory and taste sensation. Ripening marks the completion of development of a fruit and the commencement of senescence, and it is normally an irreversible event. The following sections will discuss the general nature of fruit ripening, respiratory behaviour and the involvement of the gas ethylene (C_2H_4) with these processes.

Ripening is the result of a complex of changes, many of them probably occurring independently of one another. A list of the major changes that together make up fruit ripening is given in Table 3.1. The time course of some of these changes is shown in Figure 3.3 for banana, which is a climacteric fruit. The principal difference between the climacteric tomato and banana and the non-climacteric pineapple is the presence of the respiratory peak that is characteristic of climacteric fruits. A sharp increase in respiration is shown by the increase in production of carbon dioxide or decrease in internal oxygen concentration. Research on respiration and ethylene production has gained priority in attempts to explain the mechanism of fruit ripening. Other changes occurring in climacteric and non-climacteric fruit are further characterised later in this chapter (see 'Chemical changes during maturation').

Table 3.1 Changes that may occur during the ripening of fleshy fruit

Seed maturation
Colour changes
Abscission (detachment from parent plant)
Changes in respiration rate
Changes in rate of ethylene production
Changes in tissue permeability and cellular compartmentation
Softening: changes in composition of pectic substances
Changes in carbohydrate composition
Organic acid changes
Protein changes
Production of flavour volatiles
Development of wax on skin

SOURCE Adapted from H.K. Pratt (1975) *The role of ethylene in fruit ripening. Facteurs et régulation de la maturation des fruits.* Centre National de La Recherche Scientifique: Paris, France, pp. 153–60.

Physiology of respiration

A major metabolic process taking place in harvested produce, or in any living plant product, is respiration. Respiration can be described as the oxidative breakdown of the more complex materials normally present in cells, such as starch, sugars and organic acids, into simpler molecules, such as carbon dioxide and water, with the concurrent production of energy and other molecules that can be used by the cell for synthetic reactions. Respiration can occur in the presence of oxygen (aerobic respiration) or in the absence of oxygen (anaerobic respiration, sometimes called fermentation).

The respiration rate of produce is an excellent indicator of the metabolic activity of the tissue and thus is a useful guide to the potential storage life of the produce. If the respiration rate of a fruit or vegetable is measured – as either oxygen consumed or carbon dioxide evolved – during the course of its development, maturation, ripening and senescent periods, a characteristic respiratory pattern is obtained. Respiration rate per unit weight is highest for the immature fruit or vegetable and then steadily declines with age (Figure 3.1).

A significant group of fruits that includes tomato, mango, banana and apple, shows variation from the described respiratory pattern in that these fruits undergo a pronounced increase in respiration coincident with ripening (Figure 3.1). Such an increase in respiration is known as a respiratory climacteric, and this group of fruits is classed as climacteric. The intensity and duration of the respiratory climacteric, first described in 1925 for the apple, varies widely amongst fruit species as depicted in Figure 3.4. The commencement of the respiratory climacteric coincides approximately with the attainment of maximum fruit

Figure 3.3 Physiochemical changes that occur during ripening of Cavendish banana (variety Williams). The peel colour stages indicate the change from green (stage 1) to full yellow (stage 6), and finally to a stage when skin spotting occurs (stage 7). Williams bananas take about 8 days to progress from stage 1 to stage 7 at 20°C.

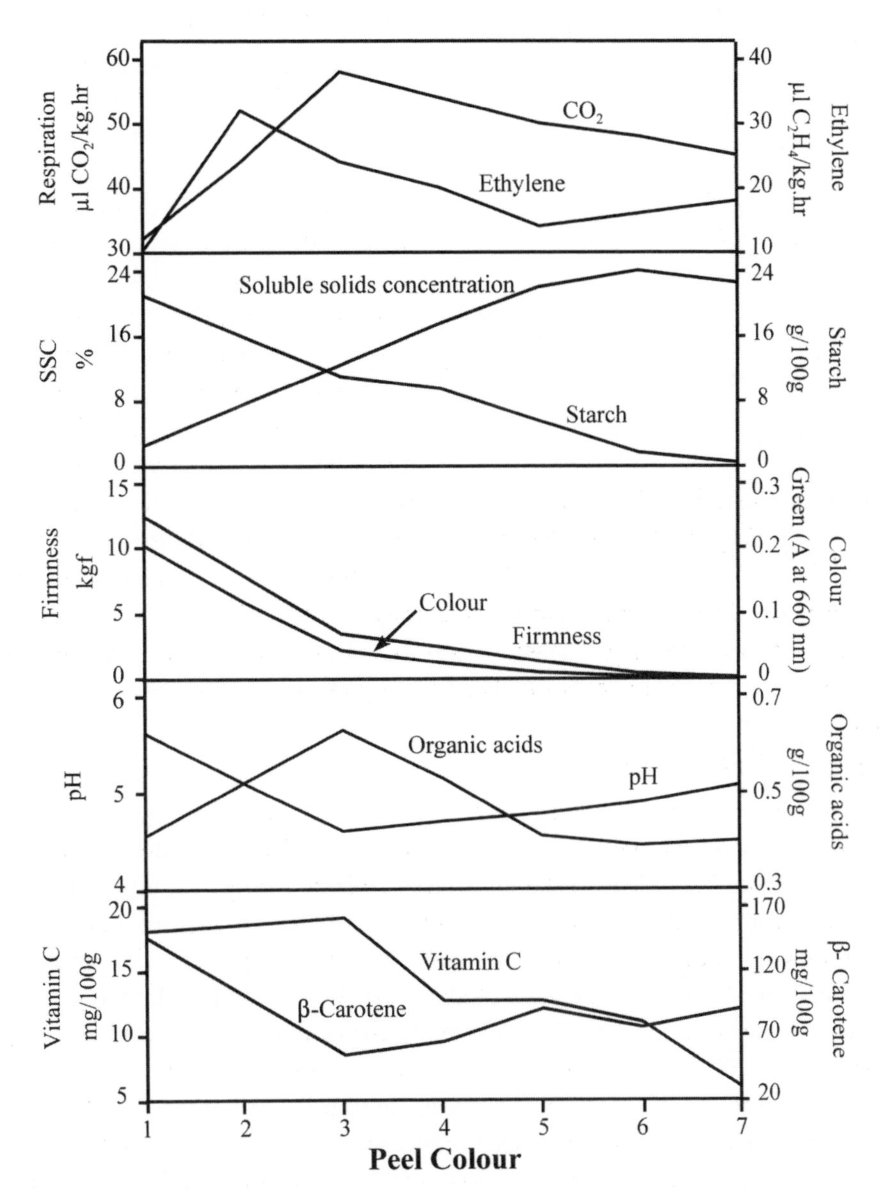

SOURCE Adapted from R.B.H. Wills, J.S.K. Lim and H. Greenfield (1984) Changes in chemical composition of 'Cavendish' banana (*Musa acuminata*) during ripening. *Journal of Food Biochemistry* 8, pp. 69–77.

Figure 3.4 Respiratory patterns of some harvested climacteric fruits

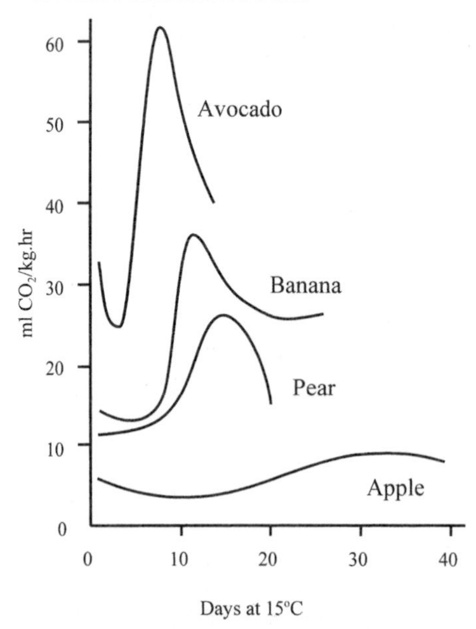

SOURCE J.B. Biale (1950) Postharvest physiology and biochemistry of fruits. *Annual Review of Plant Physiology* 1, pp. 183–206. (Used with permission.)

size (Figure 3.1), and it is during the climacteric that all the other changes characteristic of ripening occur. The respiratory climacteric, as well as the complete ripening process, may proceed while the fruit is either attached to or detached from the plant (except for avocado, which will only ripen when detached from the plant).

Fruits such as citrus, pineapple and strawberry, which do not exhibit a respiratory climacteric, are known as non-climacteric. Non-climacteric fruit exhibit most of the ripening changes, although these usually occur more slowly than those of the climacteric fruits. Table 3.2 lists some common climacteric and non-climacteric fruits. All vegetables can also be considered to have a non-climacteric type of respiratory pattern. The division of fruit into two classes on the basis of their respiratory pattern is an arbitrary classification but has stimulated considerable research into biochemical control of the respiratory climacteric.

Effect of ethylene

Climacteric and non-climacteric fruits may be further differentiated by their response to applied ethylene and by their pattern of ethylene production during ripening. It has been clearly established that all fruit produce minute quantities of ethylene during development. However, coincident with ripening, climacteric fruits produce much larger amounts of ethylene than non-climacteric fruits. This difference between the two classes of fruit is further exemplified by the internal ethylene concentration found at several stages of development and ripening (Table 3.3). The internal ethylene concentration of climacteric fruits varies widely, but that of non-climacteric fruits changes little during development and ripening. Ethylene, applied at a concentration as low as 0.1–1.0 μL/L for one day, is normally sufficient to hasten full ripening of climacteric fruit (Figure 3.5), but the magnitude of the climacteric is relatively independent of the concentration of applied ethylene. In contrast, applied ethylene merely causes a transient increase

Table 3.2 Classification of some fruits according to respiratory behaviour during ripening

Climacteric fruit	Non-climacteric fruit
Apple (*Malus domestica*)	Cherry sweet (*Prunus avium*)
Avocado (*Persea americana*)	sour (*Prunus cerasus*)
Banana (*Musa* sp.)	Cucumber (*Cucumis sativus*)
Blueberry (*Vaccinium corymbosum*)	Grape (*Vitis vinifera*)
Cherimoya (*Annona cherimola*)	Lemon (*Citrus limon*)
Fig (*Ficus carica*)	Pineapple (*Ananas comosus*)
Kiwifruit (*Actinidia deliciosa*)	Satsuma mandarin (*Citrus unshu*)
Muskmelon (*Cucumis melo*)	Strawberry (*Fragaria* sp.)
Papaya (*Carica papaya*)	Sweet orange (*Citrus sinensis*)
Passionfruit (*Passiflora edulis*)	Tamarillo (tree tomato) (*Cyphomandra betacea*)
Peach (*Prunus persica*)	
Tomato (*Lycopersicon esculentum*)	
Watermelon (*Citrullus lanatus*)	

in the respiration of non-climacteric fruits, the magnitude of the increase being dependent on the concentration of ethylene (Figure 3.5). Moreover, the rise in respiration in response to ethylene may occur more than once in non-climacteric fruits, in contrast to the single respiration increase in climacteric fruits.

Table 3.3 Internal ethylene concentrations measured in several climacteric and non-climacteric fruits

Fruit	Ethylene (μL/L)
Climacteric	
Apple	25–2500
Pear	80
Peach	0.9–20.7
Avocado	28.9–74.2
Mango	0.04–3.0
Passionfruit	466–530
Plum	0.14–0.23
Non-climacteric	
Lemon	0.11–0.17
Lime	0.30–1.96
Orange	0.13–0.32
Pineapple	0.16–0.40

SOURCE S.P. Burg and E.A. Burg (1962) The role of ethylene in fruit ripening. *Plant Physiology* 37, pp. 179–89. (Used with permission.)

Figure 3.5 Effects of applied ethylene on respiration of climacteric and non-climacteric fruits

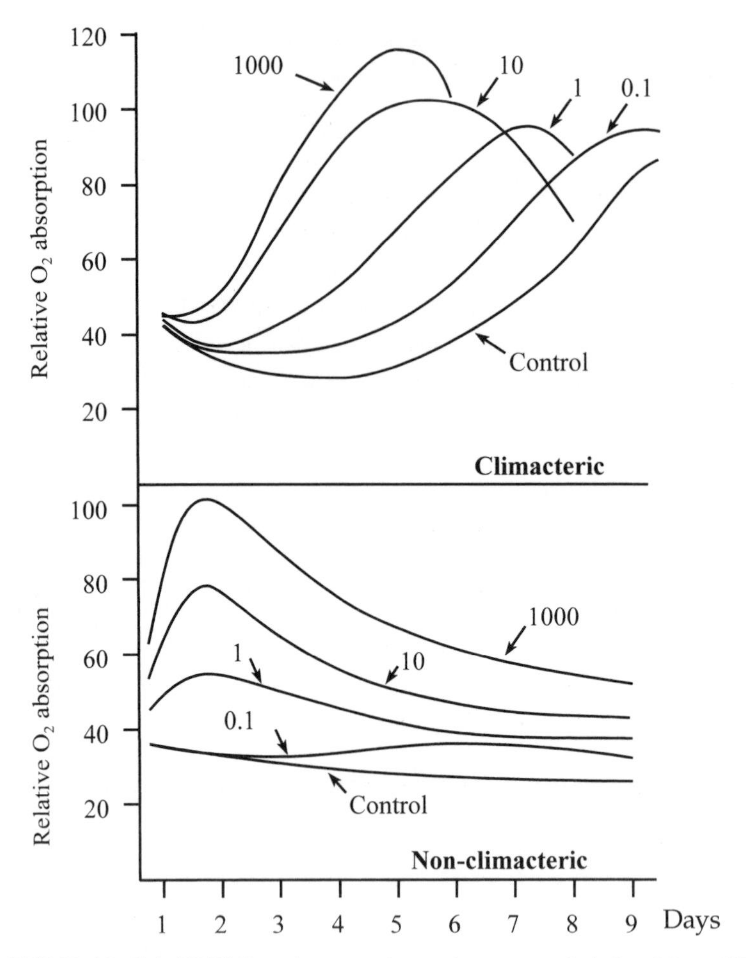

SOURCE J.B. Biale (1964) Growth, maturation, and senescence in fruits. *Science* 146, pp. 880–88. (Used with permission.)

Ethylene biosynthesis

Ethylene has been shown to be produced from methionine via a pathway that includes the intermediates S-adenosyl-methionine (SAM) and 1-aminocyclopropane-1-carboxylic acid (ACC) (Figure 3.6). The conversion of SAM to ACC by the enzyme ACC synthase is thought to be the rate-limiting step in the biosynthesis of ethylene. This enzyme is located in the cytoplasm. In higher plants, ACC can be removed by conjugation to form malonyl ACC or glutamyl ACC. Addition of ACC to preclimacteric (unripe) fruit generally results in only a small increase in ethylene evolution,

Figure 3.6 Reaction sequences in the metabolism of ethylene and its action

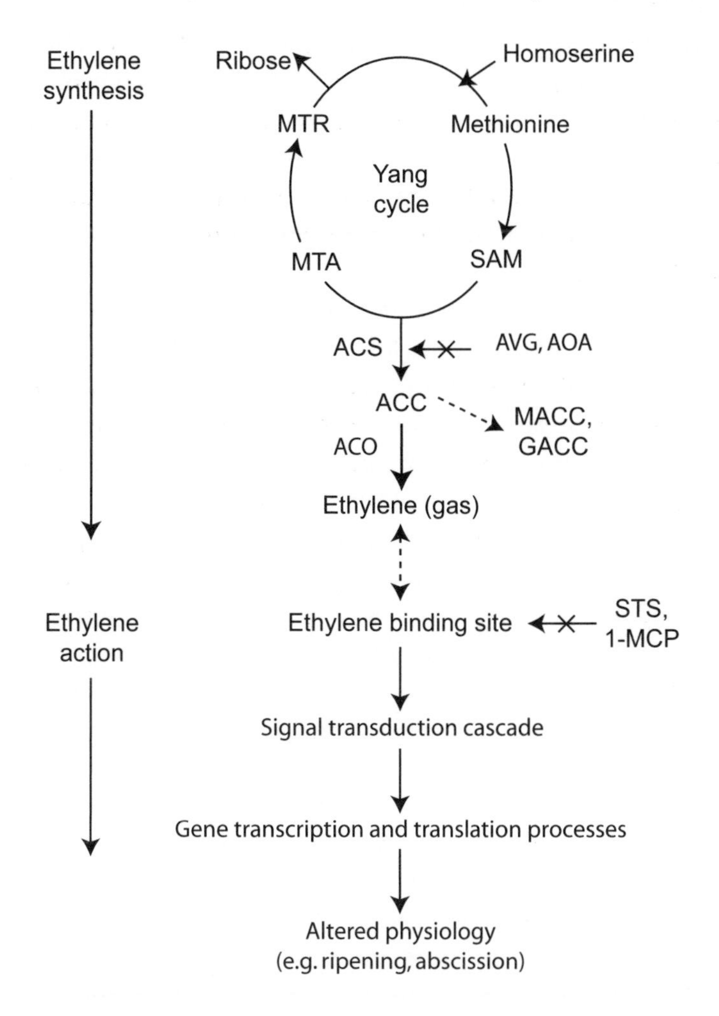

showing that another enzyme, the ethylene-forming enzyme (EFE or ACC oxidase), is required to convert ACC to ethylene. ACC oxidase is a labile enzyme, sensitive to oxygen. It is thought to be attached to the outer face of the plasmalemma. Factors that affect the activity of ACC synthase include fruit ripening, senescence, auxin, physical injuries, and chilling injury. This enzyme is strongly inhibited by aminooxyacetic acid (AOA), rhizobitoxine and its analogue and L-2-amino-4-(2-aminoethoxy)-trans-3-butenoic acid (AVG), which are known inhibitors of pyridoxal phosphate-dependent enzymes. ACC oxidase is inhibited by anaerobiosis, temperatures above 35°C and cobalt ions. Small amounts of ethylene can also be formed in plant tissues from the oxidation of lipids involving a free-radical mechanism.

Mode of action

Ethylene is a plant hormone that acts in concert with other plant hormones (auxins, gibberellins, kinins and abscisic acid) to exercise control over the fruit ripening process. Most is known about the relation of ethylene to fruit ripening because the availability of the sensitive, gas chromatographic method for measurement of ethylene has enabled detailed studies of this relationship. The relationship of the other plant hormones to ripening has not been clearly defined.

It has been proposed that two systems exist for the regulation of ethylene biosynthesis. System 1 refers to the low levels of ethylene production in immature developing fruit and vegetative tissues. The factors controlling system 1 are unknown but different forms of ACC synthase (ACS) and ACC oxidase (ACO) are thought to be involved compared to system 2. Once climacteric fruit reach a particular developmental stage there is a transition to system 2, which is responsible for producing the large amounts of ethylene necessary for the full integration of ripening. System 2 is an autocatalytic process, the produced ethylene triggering further production. Non-climacteric fruits and vegetative tissues do not have an active system 2. Treatment of developing climacteric fruits with ethylene can induce system 2 and premature ripening, but fruit treated this way has poor eating quality.

Plant hormones affect plant regulation and development by binding to specific receptors in the plant or fruit. This sets off a flow of events leading to visible responses. The ethylene receptors are in an active state in the absence of ethylene and this allows normal growth in plants and fruit to continue. When ethylene is produced naturally during ripening, following a physical stress or when it is applied artificially in a ripening room, it binds to the receptor and switches it off, leading to a series of events such as ripening or healing of injuries in some plant organs. Ethylene action can be affected by altering the amount of receptors or by interfering with the binding of ethylene to its receptors. Binding of ethylene takes place reversibly at a site containing a metal, possibly copper. The affinity of the receptor for ethylene is increased by the presence of oxygen and decreased by carbon dioxide. The occurrence of a metal-containing receptor has not been confirmed, but the proposition is supported by studies with silver ions, which inhibit the action of ethylene when applied to fruit, flowers and other tissues. The specific structural requirements for ethylene action have been demonstrated by treating tissues with analogues and antagonists of ethylene. The gaseous cyclic olefine, 1-methylcyclopropene (1-MCP) has been shown to be a highly effective inhibitor of ethylene action. 1-MCP binds irreversibly to the ethylene receptors in sensitive plant tissues and a single treatment with low concentrations for a few hours at ambient temperatures confers protection against ethylene for several days (see Chapter 6).

The pattern of changes in ethylene production rates and the internal concentrations of ethylene in relation to the onset of ripening have been observed in several climacteric fruits. In one type of fruit, ethylene concentration rises before the onset of ripening, determined as the initial respiratory increase, e.g. in tomato and honeydew melon. In the second type, ethylene does not rise before the increase in respiration, e.g. in apple and mango. In honeydew melon, the internal ethylene concentration rises from the preclimacteric level of 0.04 μL/L to 3.0 μL/L, at which concentration the fruit begins to ripen. Treatments that prevent ethylene from reaching a triggering concentration will delay ripening.

It is well known that many fruits, as they develop and mature, become more sensitive to ethylene. For some time after anthesis (flowering), young fruit can have high rates of ethylene production. Early in the life of fruit the concentration of applied ethylene required to initiate ripening is high, and the length of time to ripen is prolonged but decreases as the fruit matures (Table 3.4). The tomato is an extreme case of tolerance to ethylene. Banana and melons, in contrast, can be readily ripened with ethylene even when immature. Little is known about the factors that control the sensitivity of tissue to ethylene.

Table 3.4 Effect of maturity on the time to ripen for tomato

	Days to ripen	
Maturity at harvest (days after anthesis)	**Treated with ethylene**	**Control**
17	11	–*
25	6	–*
31	5	15
35	4	9
42	1	3

* Had failed to ripen when experiment was terminated.

NOTE Time to ripen was days between anthesis and first detectable red colour (first colour stage). Fruit was treated continuously with 1000 μL/L ethylene.

SOURCE J.M. Lyons and H.K. Pratt (1964) Effect of stage of maturity and ethylene treatment on respiration and ripening of tomato fruits. *Proceedings of the American Society of Horticultural Science* 84, pp. 491–500. (Used with permission.)

The earlier concept that an initial triggering of ripening by a single dose of ethylene is sufficient to ensure ripening is not now accepted, since application of silver ions will not only block the initiation of ripening by exogenous ethylene but also will arrest the ripening process during its progress. For example, colour development and enzyme synthesis will cease. Furthermore, some of the effects of storage under modified atmospheres (Chapter 6) are that the fruit produces reduced levels of ethylene and the development of colour and texture changes is greatly retarded, whereas the changes in sugars and acids responsible for some of the flavour changes proceed normally. It is therefore apparent that ethylene is only one of the regulatory components involved in ripening.

Ripening has long been considered to be a process of senescence, due to a breaking down of the cellular integrity of the tissue. Some ultrastructural and biochemical evidence supports this view. It is widely accepted that ripening is a genetically programmed phase in the development of plant tissue, with altered nucleic acid and protein synthesis occurring at the commencement of the respiratory climacteric resulting in new or enhanced biochemical reactions operating in a coordinated manner. Both views fit with the known degradative and synthetic capacities of fruit during ripening. In view of the ample evidence of the ability of ethylene to initiate biochemical and physiological events, it is evident that ethylene action is regulated at the level of gene expression.

Genetic control of ripening

In the climacteric fruits avocado, banana, tomato and peach, there is a marked increase in the synthesis of protein and nucleic acids (especially mRNA) preceding, and in the early stages of, the climacteric rise in respiration. There is an initial increase in certain proteins already present in unripe fruit, but later in ripening new types of proteins are synthesised. Similarly, mRNAs increase during ripening, and *in vitro* translation of these mRNAs can be shown to produce the new proteins. Application of modern molecular biology techniques such as microarrays has detected changes in gene activity as fruit such as peach undergoes the transition from the preclimacteric to climacteric stages of ripening. Similar biochemical transitions have been found in non-climacteric fruit such as grape and strawberry, although the changes are more protracted and only occur while the fruit is attached to the parent vine or plant.

Natural mutants of tomato and variants produced by genetic transformation (transgenics) that exhibit abnormal ripening behaviour have proved valuable in studying the genetic regulation of ripening. Some non-ripening or slow-ripening mutants affect the processes of ethylene synthesis, ethylene perception and/or signal transduction leading to abnormal colouring (often yellow or pale red), lack of softening or low ethylene production (see Table 3.5). These mutants have been extensively used to determine some of the fundamental processes involved in ripening, particularly softening. There are a number of hydrolase enzymes that break down the carbohydrate polymers, e.g. pectins, celluloses and hemicelluloses, responsible for the structural integrity of cell walls. Among these is the enzyme polygalacturonase (PG), which hydrolyses the α-1,4 linkage between galacturonic acid residues in pectins. Early research with the tomato mutants suggested this could be the primary enzyme responsible for softening. However, it is now known that softening is a highly complex process, involving the activities of a range of different proteins including expansins, pectic lyases and ß-galactosidases as well as hydrolases, leading to the wide array of changes in texture that occur in fruit of different species during ripening and senescence. There is a well-characterised cooperative relation between PG and pectin methylesterase in fruit such as tomato. De-esterification of pectins is required before PG can catalyse depolymerisation of pectin chains. The activity of these enzymes, the release of Ca^{2+}, which is important in cross-linking the polymer chains in the cell wall, and the consequent swelling in the middle lamella of the adjacent cell walls allows the cells to move apart. This gives the softer or melting texture of a ripe dessert peach and the reduction in crispness of senescing apples.

Table 3.5 Mutants of tomato with abnormal ripening behaviour

Mutant	Location on chromosome	Fruit phenotype compared to wild type
Ripening inhibitor *(rin)*	5	Normal growth, slowly turns pale yellow; very low ethylene production; little softening, very low PG activity; does not ripen after exposure to ethylene; high oxygen causes slight pink colour
Never ripe *(Nr)*	9	Normal growth, slowly turns orange-red; limited softening, ethylene production, PG and lycopene synthesis
Non-ripening *(nor)*	10	More extreme than *rin*; final colour is deep yellow; very low ethylene production; contains <1% of wild type PG; high NaCl causes faster ripening, deep orange colour and some softening

SOURCE Adapted from G. Hobson and D. Grierson (1993) Tomato. In G.B. Seymour, J.E. Taylor and G.A. Tucker (eds), *Biochemistry of Fruit Ripening*. Chapman & Hall: London, UK, pp. 405–42.

Biochemistry of respiration

All living organisms require a continuous supply of energy. This energy enables the organism to carry out the necessary metabolic reactions to maintain cellular organisation, to transport metabolites around the tissue and to maintain membrane permeability. In addition, a continuous supply of the organic molecules is required for synthetic reactions in cells.

Aerobic metabolism

Most of the energy required by horticultural produce is supplied by aerobic respiration, which involves the oxidative breakdown of certain organic substances stored in the tissues. Conventionally a common substrate for respiration is glucose, and, if it is completely oxidised, the overall reaction is:

$$C_6H_{12}O_6 + 6O_2 \longrightarrow 6CO_2 + 6H_2O + \text{energy}$$

Respiration is essentially the reverse of photosynthesis, by which energy derived from the sun is stored as chemical energy, mainly in the form of carbohydrates such as starch and sucrose.

Basically four processes (shown in greatly simplified form in Figure 3.7) are involved in aerobic respiration:

- Conversion of carbohydrate to six-carbon (6-C) sugar phosphates of glucose or fructose by addition of inorganic phosphate. These hexose phosphates are the true initial substrates for respiration via glycolysis, although by convention glucose is often stated for convenience.
- Glycolysis, which splits the hexose phosphates to two molecules of 3-C pyruvate, each of which is converted to 2-C acetyl Coenzyme A (acetyl-CoA). The total energy produced by converting one glucose molecule into 2 molecules of pyruvate is 8 ATP. The process occurs in the cytoplasm.
- The TCA cycle, which takes place in the mitochondria and oxidises acetyl-CoA to carbon dioxide and water. The acetyl-CoA combines with a 4-C organic acid to give a 6-C organic acid and then undergoes a series of reactions to 5-C and 4-C organic acids to give the original 4-C acid for the next round of the cycle. The two molecules of pyruvate yield 30 ATP, in addition to the 8 ATP from the glycolytic process.
- The electron transport chain, located on the inner membranes of the mitochondria, is a series of dehydrogenase enzymes, cytochrome proteins, ubiquinones and other compounds that transfer electrons from NADH and $FADH_2$, produced in the reactions of glycolysis and

the TCA cycle, resulting in the synthesis of ATP and regeneration of the oxidised forms of NAD⁺ and FAD for reuse in the respiratory reactions. This oxidative process requires molecular oxygen from the atmosphere and also produces water.

The production of carbon dioxide and the uptake of oxygen by these metabolic processes enables the rate of aerobic respiration to be measured physiologically in plant tissues.

Figure 3.7 Simplified diagram of the biochemistry of aerobic respiration in plants. Storage carbohydrate is converted to hexose phosphate and then glycolysis produces acetyl Coenzyme A, which enters the mitochondrial TCA cycle for oxidation to carbon dioxide and water, involving the electron transport chain. Energy is stored in ATP and in reduced cofactors NAD^+ or FAD. Oxygen is required.

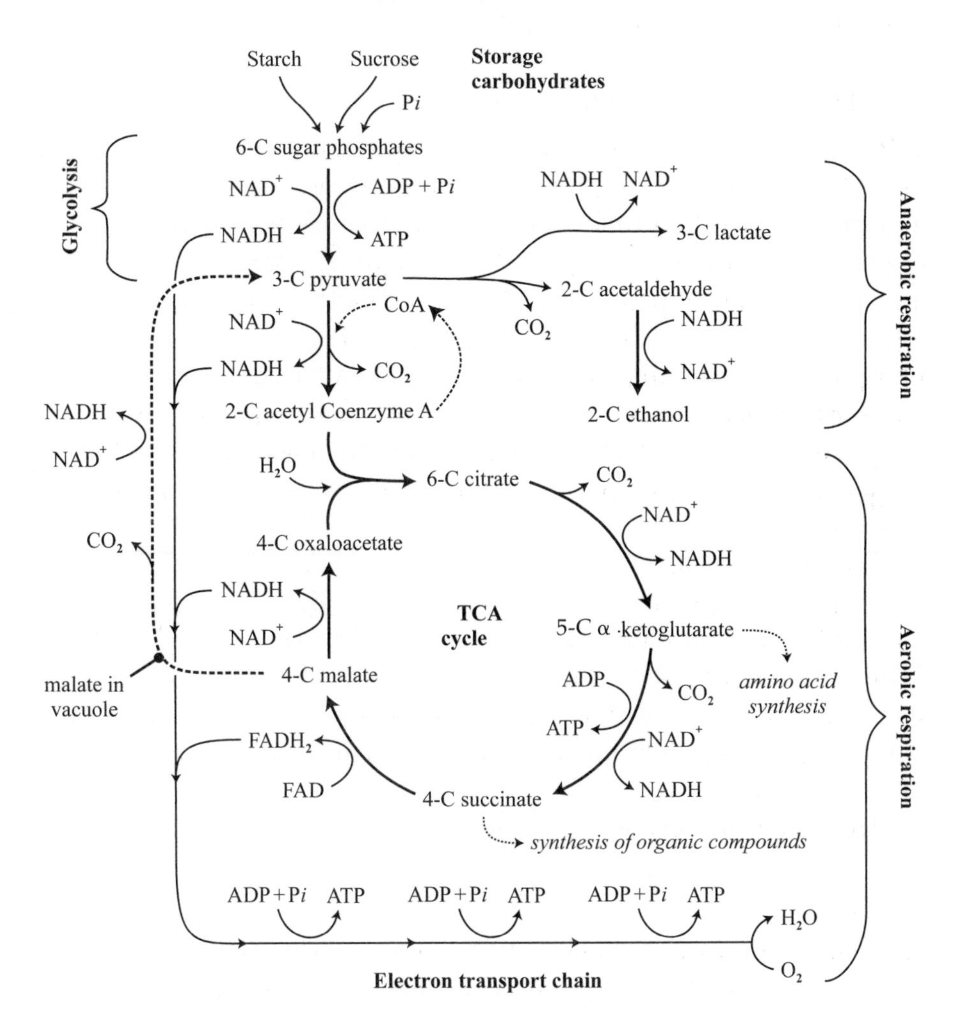

ATP is the energy currency of cells, with the energy generated during glycolysis and the TCA cycle trapped and stored in the third phosphate bond of ATP. Since oxygen is required, the process is termed oxidative phosphorylation. The energy is subsequently made available to the plant by breaking this phosphate bond in the reverse reaction:

$$ATP \longrightarrow ADP + P_i + energy$$

The resulting energy can be used in various synthetic and other metabolic reactions in the plant cells. The total chemical energy liberated during the oxidation of 1 mole of glucose is approximately 1.6 megajoules. While most of this energy is preserved within the plant system, the remainder is lost as heat. The heat generated by respiration is a major reason that refrigeration is needed during prolonged storage.

Other respiratory pathways

The oxidative pentose phosphate pathway (OPPP) converts glucose-6-phosphate to fructose-6-phosphate plus glyceraldehyde-3-phosphate and carbon dioxide through a complex cyclic reaction involving 4-, 5-, 6- and 7-carbon sugar phosphates. Although it is not considered to be a major respiratory pathway in fruits, the OPPP does provide ribose-5-phosphate for nucleotide and nucleic acid synthesis, erythrose-4-phosphate for biosynthesis of shikimic acid and aromatic amino acids, and NADPH to drive a variety of synthetic reactions. The OPPP is of greater importance in leafy vegetables and ornamentals, in which it can account for a significant proportion of tissue respiration, perhaps in the range of 10–20 per cent.

The vacuoles of many fruits and vegetables contain high concentrations of organic acids, particularly malic and citric acids, which can be used as respiratory substrates. These acids can be utilised directly by the TCA cycle in the mitochondria. Malic acid can also undergo reductive decarboxylation with the evolution of carbon dioxide and the production of NADPH or NADH and pyruvate by malic enzyme present in the cytosol or in the mitochondrion.

The complete oxidation of malate:

$$C_4H_6O_5 + 3O_2 \longrightarrow 4CO_2 + 3H_2O$$

generates more carbon dioxide than the amount of oxygen consumed, whereas the oxidation of glucose generates an equal amount of carbon dioxide for the oxygen consumed. This relationship becomes important when measuring respiration by gas exchange, in which the carbon dioxide evolved and/or oxygen consumed is measured. That is, it is possible to record different values for respiration depending on which gas is monitored. Ideally both gases should be measured simultaneously.

Anaerobic metabolism (fermentation)

Aerobic respiratory pathways are the preferred pathways, and oxygen is generally in unlimited supply. However, in various storage conditions the amount of oxygen in the atmosphere may be limited and insufficient to maintain full aerobic metabolism. Under these conditions the tissue can initiate anaerobic respiration, by which glucose is converted to pyruvate via the glycolytic pathway. But then the pyruvate is metabolised into either lactic acid or acetaldehyde and ethanol in a process termed fermentation (Figure 3.7). The oxygen concentration at which anaerobic respiration commences varies between tissues and is known as the extinction point, but is usually below 1 per cent v/v. The oxygen concentration at this point depends on several factors, such as species, cultivar, maturity and temperature, which affect the diffusion of gas to the cells. Anaerobic respiration produces much less energy per mole of glucose metabolised than aerobic pathways, but it does allow some energy to be made available to the tissue under adverse conditions. Off flavours can result from fermentation.

Metabolites for synthetic reactions

The respiratory pathways are not only used to produce energy for the tissue. Carbon skeletons are required for many synthetic reactions in the cell, and these skeletons can be removed from the TCA cycle at several points. For example, 5-C α-ketoglutarate may be converted to the amino acid glutamate, from which several other amino acids may be produced for protein synthesis; 4-C succinate may be diverted into the synthesis of various heme pigments including chlorophyll. The loss of α-ketoglutarate and succinate from the TCA cycle for synthetic reactions would eventually lead to the cycle stopping. Therefore, 4-C acids are fed into the cycle; they are produced principally by the fixation of carbon dioxide into phosphoenol-pyruvate to give 4-C oxaloacetate. Alternatively, vacuolar reserves of, for example, malate may be utilised.

Genetic control of plant metabolism

Environmental factors have long been used, in a more or less empirical fashion, to control plant metabolism; for example, by storing fruits and vegetables at low temperature and by rapidly cooling vegetables (particularly leafy ones) after harvesting. Cooling markedly reduces the rate of respiration of tissues, generally prolonging their useful life (Chapter 4). Controlled atmosphere and modified atmosphere storage of fruits and vegetables (Chapter 7), using elevated concentrations of carbon dioxide and lowered concentrations of oxygen compared to air, are also used to slow respiratory processes in produce.

More subtle manipulation of the postharvest behaviour of fruits and vegetables can be achieved by using mutant varieties or genetic manipulation to develop transgenic plants in which particular enzymes (gene products) can be either virtually eliminated or enhanced. A notable example is reducing the activity of the enzyme endopolygalacturonase, which is involved in cell wall breakdown during ripening of the tomato. Research on this enzyme resulted in the short-lived commercial production and marketing of the FLAVR SAVR™ tomato. This transgenic tomato remains firmer longer during ripening and can thus be left on the plant longer before picking, with the object of enhancing flavour. Transgenic fruits with ACC deaminase, antisense ACC synthase, ACC oxidase and lowered polyphenoloxidase have also been produced. The first three modifications reduce ethylene production and slow ripening, while lowered polyphenoloxidase activity reduces browning of damaged tissue. Thus far no transgenic fruits are being grown commercially because of consumer concerns about eating genetically manipulated produce.

Apart from consumer concerns, genetic techniques need to be approached with some caution since plant metabolism is very adaptable; there are usually alternate pathways by which a given metabolic product or intermediary metabolite can be produced. This plasticity of metabolism means plants are able to modify their performance under a wide variety of conditions, but it makes the task of the genetic engineer more difficult. This is borne out by studies in which particular enzyme levels have been modified in plants by genetic manipulation in order to study metabolic control. It was thought that a small number of key enzymes in each pathway regulated the flow of carbon through a particular pathway, but virtual removal of a particular enzymic activity by genetic manipulation was found to have little effect in a number of cases, so a combination of the plasticity of metabolism (mentioned above) and much more dispersed regulatory control mechanisms must be involved.

Chemical changes during maturation

At some stage during the growth and development of fruit and vegetables, the produce is recognised by consumers as having attained optimum eating condition. This desirable quality is not associated with any universal change, but is attained in various ways in different tissues (see Chapter 10).

Fruit

Climacteric fruits generally reach the fully ripe stage after the respiratory climacteric. However, other events initiated by ethylene are what consumers associate with ripening.

Colour

Colour change is the most obvious signal; it occurs in many fruits and is often the major criterion used by consumers to determine whether fruit is ripe or unripe. The most common change is the loss of green colour. With a few exceptions, e.g. avocado, kiwifruit and Granny Smith apple, climacteric fruits show rapid loss of green colour on ripening. Many non-climacteric fruits also exhibit a marked loss of green colour with attainment of optimum eating quality, for example citrus fruit in temperate climates (but not in tropical climates). The green colour is due to the presence of chlorophyll, which is a magnesium-organic complex. The loss of green colour is due to degradation of the chlorophyll structure. The principal agents responsible for this degradation are pH changes (mainly due to leakage of organic acids from the vacuoles), oxidative systems and chlorophyllases (Figure 3.8). Loss of colour depends on one or all of these factors acting in sequence to destroy the chlorophyll structure.

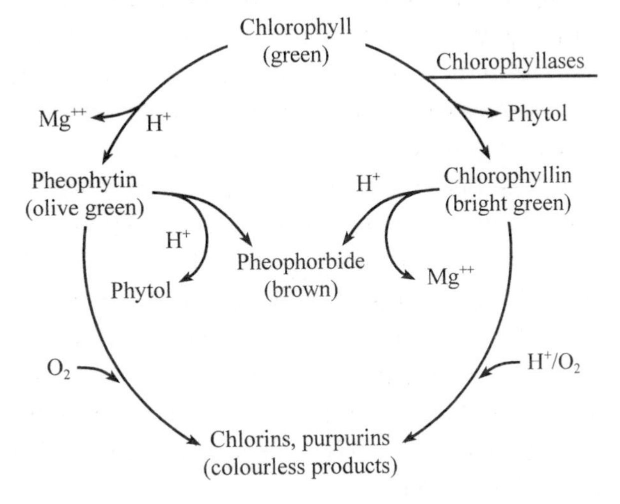

Figure 3.8 Some pathways for the degradation of chlorophyll

The disappearance of chlorophyll is often associated with the synthesis and/or the revelation of pigments ranging from yellow to red. Many of these pigments are carotenoids, which are unsaturated hydrocarbons with generally 40 carbon atoms and sometimes one or more oxygen atoms in the molecule. Carotenoids are stable compounds and remain intact in the tissue even when extensive senescence has occurred. Carotenoids may be synthesised during the development stages on the plant, but they are masked by the presence of chlorophyll. Following the degradation of chlorophyll, the carotenoid pigments become visible. With other tissues, carotenoid synthesis occurs concurrently with chlorophyll degradation. Banana peel is an example of the former system and tomato of the latter.

Anthocyanins provide many of the red-purple colours of fruit and vegetables and flowers. Anthocyanins are water-soluble phenolic glucosides that can be found in the cell vacuoles of fruit and vegetables such as beetroot, but are often in the epidermal layers, as with apple and grape. They produce strong colours, which often mask carotenoids and chlorophyll.

Carbohydrates

The largest quantitative change associated with ripening is usually the breakdown of carbohydrate polymers; particularly frequent is the near-total conversion of starch to sugars. This affects both the taste and the texture of produce. The increase in sugar renders fruit much sweeter and, therefore, more acceptable. Even with non-climacteric fruits, the accumulation of sugar is associated with the development of optimum eating quality, although the sugar may be derived from sap imported into the fruit rather than from the breakdown of starch reserves within the fruit.

The breakdown of polymeric carbohydrates, especially pectic substances and hemicelluloses, weakens cell walls and the cohesive forces binding cells together. In the initial stages, the texture becomes more palatable, but eventually the plant structures disintegrate. Protopectin is the insoluble parent form of pectic substances. In addition to being a large polymer, it is cross-linked to other polymer chains with calcium bridges and is bound to other sugars and phosphate derivatives to form an extremely large polymer. During ripening and maturation, protopectin is gradually broken down to lower molecular weight fractions, which are more soluble in water. The rate of degradation of pectic substances is directly correlated with the rate of softening of fruit.

Organic acids

Usually organic acids decline during ripening as they are respired or converted to sugars. Acids can be considered as a reserve source of energy to the fruit and would, therefore, be expected to decline during the greater metabolic activity that occurs on ripening. There are exceptions, such as banana, where the highest level is attained at the fully ripe stage, but the level is not high at any stage of development compared to other produce.

Aroma

Aroma plays an important part in the development of optimal eating quality in most fruit. It is due to the synthesis of many volatile organic compounds (often known merely as volatiles) during the ripening phase. The total amount of carbon involved in the synthesis of volatiles is less than 1 per cent of that expelled as carbon dioxide. The major volatile formed is ethylene, which accounts for about 50–75 per cent of the total carbon in the volatiles; ethylene does not contribute to typical fruit aromas. The amount of aroma compounds is therefore extremely small. Chapter 2 discussed the nature of the compounds formed. Non-climacteric fruits also produce volatiles during the development of optimum eating quality. These fruits do not synthesise compounds as aromatic as those in climacteric fruit; nevertheless, the volatiles produced are still important in consumer appreciation.

Vegetables

Vegetables generally show no sudden increase in metabolic activity that parallels the onset of the climacteric in fruit, unless sprouting or regrowth is initiated. The process of germination is sometimes deliberately applied to some seeds, for example, mung bean, and the sprouted product is the marketed vegetable. Apart from obvious anatomical changes during sprouting, considerable compositional changes occur. The sugar level increases markedly as the result of the rapid conversion of fats or starch. From a nutritional point of view, the increase in vitamin C in sprouted seeds can be valuable in diets that have a marginal vitamin C intake.

Vegetables can be divided into three main groups: seeds and pods; flowers, buds, stems and leaves; bulbs, roots and tubers. Some fruits are also consumed as vegetables; they may be either ripe (tomato, eggplant) or immature (zucchini, cucumber, okra).

Seeds and pods, if harvested fully mature, as is the practice with cereals, have low metabolic rates because of their low water content. In contrast, all seeds consumed as fresh vegetables, for example sweet corn, have high levels of metabolic activity, because they are harvested at an immature stage, often with the inclusion of non-seed material, for example the bean pod (pericarp). Eating quality is determined by flavour and texture, not by physiological age. Generally seeds are sweeter and more tender at an immature stage. With advancing maturity, the sugars are converted to starch, with a resultant loss of sweetness; the water content also decreases and the amount of fibrous material increases. Seeds for consumption as fresh produce are harvested when the water content is about 70 per cent; in contrast, dormant seeds are harvested at less than 15 per cent water.

Edible flowers, buds, stems and leaves vary greatly in metabolic activity and hence in their rate of deterioration. Stems and leaves often senesce rapidly and so lose their attractiveness and nutritive value. The most visible sign of senescence is generally degreening, resulting in yellowing due to underlying carotenoid pigments. Texture is often the dominant characteristic that determines both the harvest date and quality, since loss of turgor through water loss causes a loss of texture. The natural flavour is often of less importance than texture, as many of these vegetables are cooked and seasoned with salt or spices. Growth processes such as cell division and expansion, and protein and carbohydrate synthesis usually cease upon harvest and the metabolism goes into a catabolic or degradative mode.

Bulbs, roots and tubers are storage organs that contain the food reserves required when growth of the plant resumes, and often some are retained for the purpose of propagation. When harvested, their metabolic rate is low and, under appropriate storage conditions, their dormancy can be prolonged. The biochemistry of these storage organs is geared to a slow metabolic rate, designed to provide the low levels of energy required to maintain life in the cells of these tissues during dormancy. Postharvest management aims to maintain the produce in its dormant state.

Ornamentals

Ornamentals can be grouped into cut flowers, cut foliage and pot plants. Others, of course, are used in gardens and urban landscaping, but these are beyond the scope of this book. Flowers usually have high rates of respiration through glycolysis of sugar translocated from the leaves. For cut flowers, therefore, the so-called preservative solutions used by florists often contain sucrose as a carbohydrate source to help maintain the respiration rate and thus extend storage life.

Pigments

The main pigments in ornamental foliage are similar to those in fruit and vegetables. Chlorophyll is the principle green foliage pigment and carotenoids constitute many of the yellow, orange and red pigments. However, the anthocyanins and related compounds are responsible for red, purple and blue colours in most flowers. The observed colours are largely related to the pH of the flower sap. At more acid pH levels (below 7) anthocyanins are red, whereas above pH 7 they tend to be blue. This gives rise to the phenomenon in roses known as 'blueing', where a shift from red to blue colouration occurs with aging. This is due to the depletion of sugars as a respiratory substrate and the switch to catabolism of proteins; the subsequent release of free amino groups results in a more alkaline pH in the cell sap. The 'colour' white is due to a total reflectance of the visible spectrum and often results from the presence of highly aerated tissues, as in some flowers. The variegated leaves in some ornamentals are due to areas being devoid of chloroplasts, and these areas may be white or orange/ yellow, the latter colour being due to the carotenoid pigments.

Ethylene

Flowers tend to have a short postharvest (vase) life. Sensitivity to ethylene is often cited as a major contributing factor for this. Most ornamentals should be regarded as non-climacteric, although some produce a distinct ethylene peak and respiratory climacteric. There will therefore be differential responses to ethylene similar to those seen in fruit and vegetables. Thus some flowers, such as the non-climacteric delphinium, are highly sensitive to ethylene, while the climacteric carnation is relatively tolerant. The major effect of ethylene is induced abscission. The short postharvest life of cut flowers is, however, confounded by limited carbohydrate reserves and an overall rapid rate of metabolism and development.

4 Effects of Temperature

Temperature is the single most important factor in maintaining the postharvest quality of fruits, vegetables and ornamentals. Temperature responses of harvested horticultural produce can be generally classed as:

- normal intermediate temperature range effects
- adverse low temperature effects, and
- adverse high temperature effects (Figure 4.1).

In the case of adverse low temperature effects, there is a clear distinction between freezing and chilling injuries. For specific disorders and further details on the nature of low and high temperature effects, see Chapters 8 and 13, respectively.

Normal intermediate temperature effects

Harvested produce is ideally transported and stored under reduced temperatures likely to maximise longevity. However, the effect of reducing temperature on maintaining produce quality is not uniform over the normal intermediate or physiological temperature range (0–30°C for non-chilling-sensitive produce, 7.5–30°C for moderately chilling-sensitive produce, and 13–30°C for chilling-sensitive produce). Only a small improvement in storage life, the time that produce can be held in an acceptable condition after harvest, is achieved by small reductions in temperature at the upper end of the temperature range. In contrast, much larger improvements are obtained

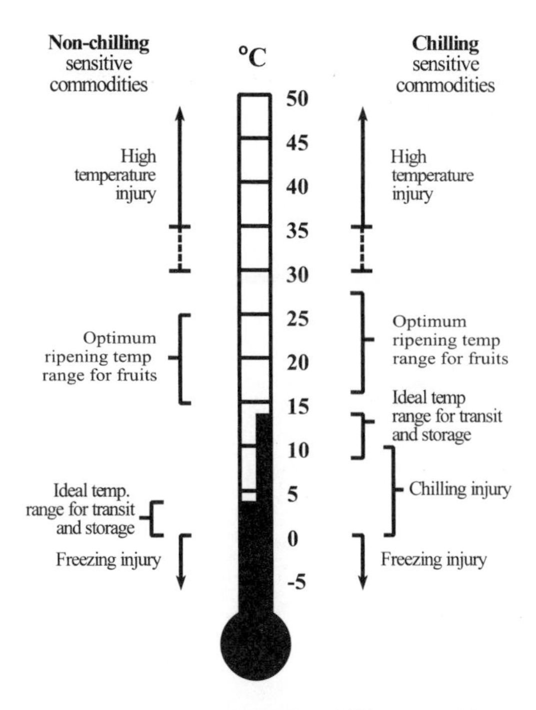

Figure 4.1 Responses of non-chilling-sensitive and chilling-sensitive produce to temperature

by similarly small reductions at lower temperatures (Figure 4.2), where even a change in temperature of 1°C can have a significant effect. The greatest reduction in processes associated with deterioration (e.g. respiration, change in texture, loss of vitamin C), and thus the best quality maintenance, will be obtained if produce is held just above its freezing point, or just above its chilling threshold temperature in the case of chilling-sensitive produce.

Metabolism (e.g. respiration) in fruit, vegetables and ornamentals involves many enzymic reactions. The rate of these reactions, within the physiological temperature range, generally increases exponentially with increase in temperature. This relationship may be described mathematically by use of the temperature quotient or Q_{10}. Van 't Hoff, a Dutch chemist, determined that the rate of a chemical reaction approximately doubles for each 10°C rise in temperature. This relationship is, however, of limited use as the Q_{10} for many biological processes does not remain constant over the physiological range; for example, the Q_{10} for respiration can have a value of between 2 and 8 across the physiological temperature range.

Lowering the temperature of both climacteric and non-climacteric produce lowers their rate of deterioration. However, in the case of climacteric fruit, low temperature can also be used to achieve a delay in the onset of ripening. The effect of decreased temperature on ripening follows an exponential relationship similar to that shown in Figure 4.2. Lowering temperature not only reduces production of ethylene, but also the rate of response of the tissues to ethylene. Thus, at lower temperature, longer exposure to a given concentration of ethylene is required to initiate ripening or enhance senescence.

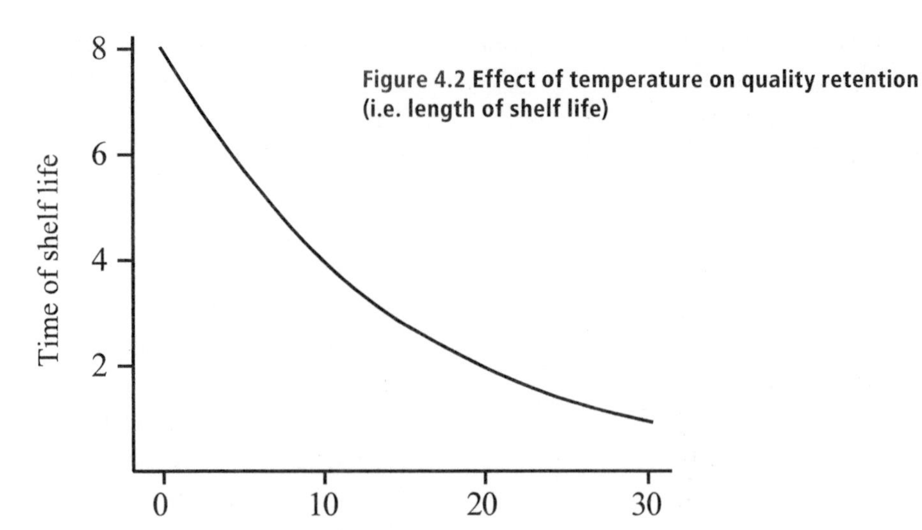

Figure 4.2 Effect of temperature on quality retention (i.e. length of shelf life)

Normal ripening occurs only within a particular range of temperature (commonly 10–30°C), although some fruit, for example some pear cultivars, will ripen slowly but satisfactorily at temperatures below 10°C. The best quality in fruit, however, generally develops at a ripening temperature of 20–23°C.

Provided that a fruit is not sensitive to chilling, maximum storage life can be achieved at temperatures below the ripening range. For example, Williams Bon Chretien (or Bartlett) pears will not ripen at temperatures below about 12°C, and maximum storage life is obtained by storage at –1°C followed by removal to temperatures greater than 12°C when ripening is desired. However, if the pears are held too long at low, non-ripening temperatures, they will fail to ripen normally after removal to ripening temperatures.

For chilling-sensitive produce, the best retention of quality is obtained just above the chilling threshold temperature, whereas for non-chilling-sensitive produce storage and handling just above the freezing temperature is optimal. In addition to delaying fruit ripening and senescence processes, these basic principles of temperature management can be applied to inhibit all manner of development processes in fresh produce; for example, opening of cut flowers, toughening of asparagus and loss of sweetness in peas.

Adverse low temperature effects

Produce may be exposed to undesirably low temperatures as a result of, for example, transport in cold climatic regions, incorrect thermostat settings in storage rooms, and exposure to sub-zero coolant (e.g. dry ice). Freezing injury occurs at temperatures of 0°C or below, and involves inter- and/or intracellular ice formation. The precise temperature at which freezing occurs depends upon the concentration of solutes in the tissue, with freezing point being lowered further with increasing osmotic concentration (i.e. freezing point depression). For example, lettuce freezes at just below 0°C (about –0.2°C), while mature grape berries, which have a very high sugar content (about 14% of fresh weight), only freeze below –2.0°C.

Freezing of tissue water initiates desiccation and causes osmotic stress to cellular structures, such as membranes, and cell constituents, such as proteins, because solvent water is lost to support ice crystal growth. In addition, expansion of the water upon freezing, especially intracellular ice formation, can cause considerable physical disruption to cell structure. Upon thawing, affected tissue usually cannot resume its normal metabolism or regain normal texture. Adversely affected freeze-thawed tissue is flaccid and/or water-soaked. However, some produce, such as cabbage, onions and some pear cultivars, may be thawed without detriment. Normal form and function can be regained if ice crystal damage is minimal and if the rate of temperature rise is sufficiently slow to allow orderly redistribution of water and for intracellular compartments to reform.

Chilling injury of susceptible commodities occurs at low temperatures that are above the freezing point of the produce. Injury is the result of imbalanced metabolism and loss of cellular compartmentation at sub-optimal temperatures. Factors that influence susceptibility to chilling injury are discussed in Chapter 8, as are the myriad chilling injury symptoms. Subtropical and tropical commodities are especially chilling sensitive, with chilling thresholds for tropical produce being around 13°C.

Manifestation of chilling injury is a function of time and temperature. A short period of storage just below the chilling threshold temperature may not result in chilling injury symptoms in specific produce. However, longer exposure will result in irreversible damage, the extent of damage increasing with increasing duration. Conversely, the development of chilling injury symptoms will be more severe in chilling-susceptible produce held for even a short period well below the chilling threshold temperature.

Adverse high temperature effects

Unusually high temperatures are associated with 'insults' such as exposure of harvested produce to direct sunlight, hot ambient air, and heat treatments for pest eradication (e.g. hot water dips; vapour and dry heat treatments). The activity of enzymes in fruit, vegetables and ornamentals declines at temperatures above 30°C. At certain temperatures, specific enzymes become inactive (denature); many are still active at 35°C but most are inactive by 40°C.

Continuous exposure of some climacteric fruit to temperatures around 30°C allows the flesh to ripen but inhibits fruit colouration. For example, the peel of Cavendish banana (cultivars Valery and Williams) remains green, and in tomato the lycopene (red pigment) accumulation is inhibited during ripening at elevated temperatures. When produce is held above 35°C, metabolism becomes abnormal and results in a breakdown of membrane integrity and structure, with disruption of cellular organisation and rapid deterioration of the produce. The changes are often characterised by a general loss of pigments, and the tissues may develop a watery or translucent appearance. This condition in banana and tomato is often referred to as 'boiled'.

Pest activity interaction with temperature

Although, microbial and insect pest activities are considered in Chapters 9 and 11, respectively, it is important to note here that greatly diminished rates of pest population growth and development can be an additional benefit of good temperature management. For example, microbial growth is often the visible symptom of wastage when the root cause is poor temperature management. Bacterial and fungal organisms exploit tissue deterioration (e.g. leakage of cellular contents) associated with adversely low or high temperatures, or simply with ongoing ripening and senescence processes in the normal physiological temperature range. Similarly, the rate of insect regeneration can be slowed at low temperatures. Accordingly, there is another obvious benefit in optimising temperature management: to minimise reliance on chemical control measures.

Sugar–starch balance

Storing some vegetables, including potato, sweet potato, green peas and sweet corn, at low temperatures can alter the starch–sugar balance in the produce. At any temperature, starch and sugar are in dynamic equilibrium, and some sugar is degraded to carbon dioxide during respiration:

$$\text{starch} \longleftrightarrow \text{sugar} \longrightarrow CO_2$$

At ambient temperatures, the starch–sugar balance in potato and sweet potato is heavily biased towards accumulation of starch. When these vegetables are stored at reduced temperatures, the rate of respiration and the conversion of sugar to starch decreases. The critical temperature at which accumulation of sugar commences depends on the commodity (it is about 10°C for potato and 15°C for sweet potato). Accumulation of sugar is profoundly undesirable in many starchy vegetables. Potato with a high sugar content has poor texture and a sweet taste when boiled; when fried, excessive browning occurs due to caramelisation and reactions between amino acids and sugars (Maillard reaction). The accumulation of sugar in potato stored at low temperatures can be largely reversed by raising the storage temperature to 10°C or above. Although it is widely accepted that the sugar level returns to nearly normal during one week at 15–20°C, experience has shown that the decrease in sugar level may occur at a much slower rate, especially after prolonged storage at low temperatures.

In other vegetables, such as sweet corn and peas, a high sugar content is desired. These vegetables are harvested immature when the sugar content is highest, and rapid storage at quite low temperatures is necessary to retard conversion of sugar to starch.

Storage life

There is no one ideal temperature for the storage of all horticultural commodities since their responses to temperature vary widely. Important physical processes such as transpiration and physiological reactions like chilling injury must be taken into account, along with the required duration of storage. In fruits, vegetables and ornamentals not susceptible to chilling injury, maximum storage life can be obtained at temperatures close to but above the freezing point of the tissue. For chilling-sensitive produce, there are also potential immediate advantages of applying low temperature during storage and distribution. These factors are considered in more detail in Chapter 13.

Cooling of produce

The object of cool storage is to slow deterioration without predisposing the commodity to abnormal ripening or other undesirable changes, thereby maintaining it in a condition acceptable to the consumer for as long as possible. The low temperature storage of horticultural produce is more demanding than for other foods. In addition to having sufficient refrigeration capacity to cool the produce to the required temperature, provision must be made for continuous removal of the heat of respiration (vital heat), maintenance of high RH, and, in some cases, ventilation or atmosphere control. Cool stores for fresh horticultural produce are generally required to operate within relatively close temperature limits (e.g. ±1°C), both in space (i.e. throughout the room) and time (i.e. constantly), in order to maximise storage life, avoid freezing, minimise desiccation and avoid gas injury (e.g. high carbon dioxide, low oxygen) to produce.

The temperature of horticultural produce at harvest is close to that of ambient air and can be in the order of 40°C for produce held in direct sunlight. At such temperatures respiration rates are extremely high, and postharvest life will be greatly reduced. It is often good practice to harvest early in the morning to take advantage of the lower produce temperatures generally prevailing at this time. Night harvesting has been investigated in certain situations, although the practice has not been sustained for reasons that include cost and inconvenience. Early morning harvesting may not be feasible for larger growers, and morning temperatures in tropical areas may still be relatively high.

The quicker the temperature of produce is reduced to the optimum storage temperature, the longer its storage life. Rapid or fast cooling after harvest is generally referred to as 'precooling', and particularly benefits highly perishable (e.g. raspberries) and/or rapidly developing (e.g. asparagus) commodities. Special facilities for the fast cooling of produce after harvest are often essential because refrigerated transport spaces, such as refrigerated ships' holds, land vehicles and shipping containers, are generally not designed to remove field heat or to allow for adequate cold air circulation within and/or around individual containers. Rather, transport units are designed to maintain precooled produce at the selected carriage temperature. It is now common for maximum acceptable loading temperatures for perishable produce to be closely controlled in order to help ensure that produce will arrive at its destination in good condition.

The term precooling is loosely applied, such that it encompasses any cooling treatment given to produce before shipment, storage or processing. A stricter definition of precooling would include only those methods by which the produce is cooled rapidly, certainly within 24 hours of harvest. No legal definition of precooling has been established, but the definition must be sufficiently broad and flexible to embrace the cooling requirements of the various commodities in relation to their required postharvest life.

The method of cooling selected will depend greatly on the anticipated storage life of the commodity. Rapidly respiring commodities, which have a short postharvest life, need to be quickly cooled soon after harvest. Commodities that have a longer postharvest life generally do not need to be cooled quite so rapidly, but nonetheless should be cooled as soon as possible. Commodities that are susceptible to chilling injury should be cooled according to their individual requirements. Generic temperature recommendations for non-chilling-sensitive, moderately chilling-sensitive and highly chilling-sensitive commodity groupings are 0, 5–7.5, and 13–15°C respectively. Selection of the most appropriate precooling method depends on three main factors: the temperature of the produce at harvest, the physiology of the produce, and the desired postharvest life. Produce that is to be 'cured' (Chapter 11) at temperatures above those required for extended storage is not normally precooled (e.g. potato, yam and sweet potato).

Cooling rates

The rate of cooling of produce is dependent primarily upon the:

- rate of heat transfer from the produce to the cooling medium, which is especially influenced by the rate of flow of the cooling medium around or through the containers of produce and the extent of contact between the two;
- difference in temperature between the produce and the cooling medium;
- nature of the cooling medium;
- thermal conductivity of the produce.

When warm produce is exposed to cold air kept at a constant temperature by refrigeration, the rate of cooling (°C/h) is not constant, but diminishes exponentially as the temperature difference (driving force) between produce and air falls (Newton's Law) (Figure 4.3). Because the rate of cooling varies over time, two single parameters have been adopted to describe the cooling process. The cooling coefficient (C) is defined as the ratio of the change in product temperature per unit time (R; °C/h) at any moment to the difference in temperature between the product and the coolant (dT; °C) at the same moment:

$$C = R/dT$$

The 'half' and 'seven-eighths' cooling times are the times required to reduce the temperature difference between the product and the cooling medium by one half (Z) or by seven-eighths (S), respectively. Theoretically, Z and S (which is equivalent to three half-cooling times) are independent of the initial product temperature, and remain constant throughout the cooling period. S is of more practical use in commercial cooling operations, because the temperature of the produce at the seven-eighths cooling time is acceptably close to the required storage or transport temperature. In systems where the cooling rate is rapid, the temperature change in the interior of produce lags considerably behind the change in surface temperature. This is particularly true for bulky products, such as melons. In such cases, the limiting factor is the rate of heat conduction to the surface of the product. This positional lag effect can alter the relative difference in time between S and Z, such that S may range from 2Z to 3Z. Mathematically 'seven-eighths' cooling is expressed as the logarithmic equation:

$$S = \ln(8j)/C$$

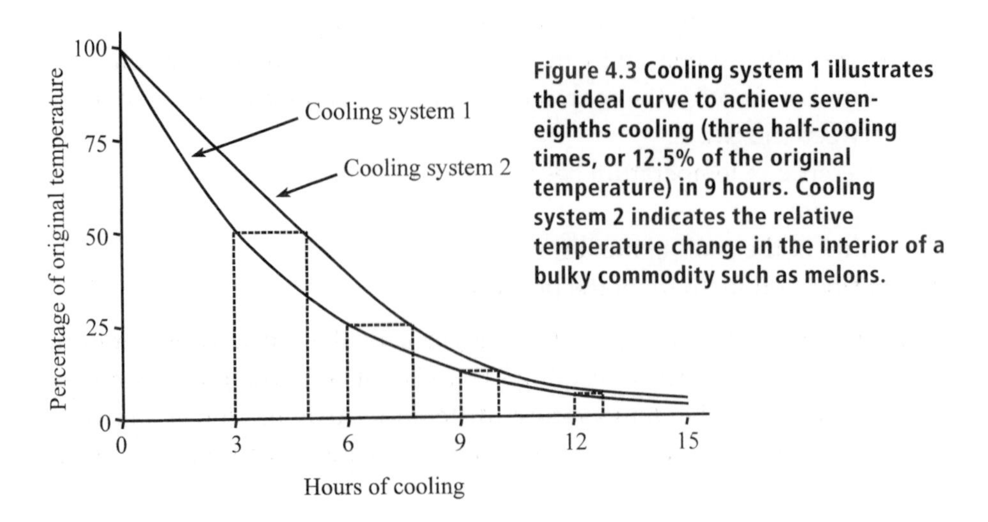

Figure 4.3 **Cooling system 1 illustrates the ideal curve to achieve seven-eighths cooling (three half-cooling times, or 12.5% of the original temperature) in 9 hours. Cooling system 2 indicates the relative temperature change in the interior of a bulky commodity such as melons.**

where j is the lag factor, which may vary from 1 to 2 at the centre of the cooling objects, and C is the cooling coefficient, a negative value.

The cooling method, type of package, and the way the packages are stacked will all influence the rate of cooling of produce. The influence of such factors on the Z value is shown in Table 4.1.

Table 4.1 Half-cooling times for apples

Cooling method		Z (h)
Packed in 18 kg box placed in:		
	Conventional cool room	12
	Air tunnel of velocity 200–400 m/min	4
Single fruit placed in air tunnel of:		
	Velocity 40 m/min	1.25
	Velocity 400 m/min	0.5
Hydrocooling of loose fruit		0.33

SOURCE E.G. Hall (1972) Precooling and container shipping of citrus fruits. *CSIRO Food Research Quarterly* 32, pp. 1–10. (Used with permission.)

Methods of cooling

Produce may be cooled by means of cold air (room cooling; forced-air or pressure cooling), cold water (hydrocooling), ice and evaporation of water (evaporative cooling; vacuum cooling). Fruit is normally cooled with cold air, although stone fruit benefits from hydrocooling. Any of the cooling methods may be used for vegetables, depending upon the structure and physiology of the specific commodities, and upon market requirements and expectations. Cut flowers and foliage are usually forced-air cooled, although they may also be vacuum cooled if a small amount of water loss is acceptable.

Room cooling

Probably the most common precooling technique is room cooling, whereby produce in containers (wooden, fibreboard, plastic), bulk containers or various other packages (e.g. mesh bags) is exposed to cold air in a normal cool store. For adequate cooling, air velocities around the packages should be at least 60 m/min. Produce may be cooled and stored in the same place, thereby minimising rehandling. This can be feasible if produce is supplied over time so that only small demands are placed on the room refrigeration system, which is relatively inefficient for cooling compared to specialised precooling systems. However, in addition to being a slow method of cooling, room cooling has the disadvantage that comparatively more space is required, as produce must be widely spaced to maximise cooling.

Pressure (forced-air) cooling

The rate of cooling with cold air is significantly increased if the heat transfer surface (i.e. contact area) is enlarged from that of the package to the total surface area of the produce within the package. By forcing air through packages and around each item of produce, pressure cooling can cool produce in about one-quarter to one-tenth the time required for room cooling.

Most commonly, pressure cooling involves passing cold air down an induced pressure difference (gradient) past initially warm produce in specially vented containers (Figure 4.4). The pressure differential is induced by fans that circulate cold air through the produce and packaging, which constitute the resistance to air flow. Pressure differentials between opposite faces of packages range from barely measurable to about 250 Pa (25 mm water head), and air-flow rates can vary between 0.1 and 2.0 L/sec.kg. Within limits, the speed of cooling can be adjusted by varying the rate of air flow. Refrigeration requirements for pressure cooling are often overestimated because of a lack of understanding of the factors that limit the rate of heat loss from the cooling produce (cooling system 2, Figure 4.3). Such overestimation of refrigeration requirements increases the initial capital cost of a cooling plant unnecessarily. The thermal properties, S and j, of stacks of packaged produce should be taken into account when calculating refrigeration capacity.

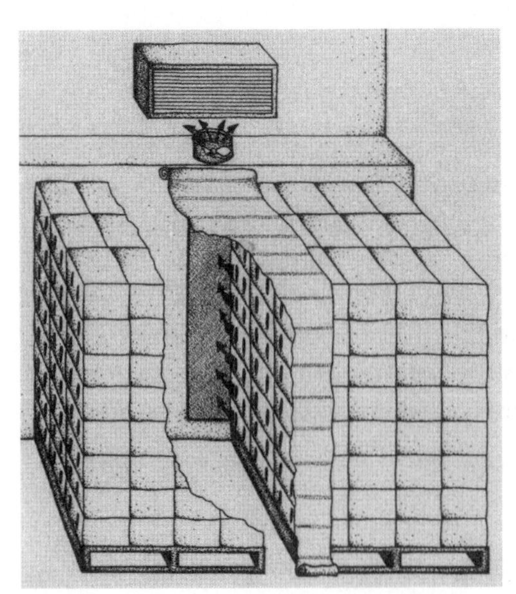

Figure 4.4 Air flow in forced-air (pressure) cooling. Packages and containers must be vented (at least 5% of exposed surface area) and stacked so that the cooling air is forced to flow through the containers and around the produce to return to the exhaust fans. A slight pressure drop occurs across the produce as air exhausts from the space between the stacks of produce.

Maintaining high RH during pressure cooling is generally not considered critically important because the process is relatively rapid. However, this generalisation applies mostly to commodities with a low surface area to volume ratio, such as fruit and fruit vegetables. High-humidity pressure-cooling systems (e.g. air-wash refrigeration systems) have been developed for use with commodities of high surface area to volume ratio, like leafy vegetables and cut flowers. Once a commodity has been adequately cooled, air velocity should be reduced or the produce transferred to a normal cool room to reduce the risk of desiccation.

Figure 4.5 Hydrocooling unit designed to remove field heat from harvested stone fruit in bulk plastic bins. The bins travel slowly through a high-volume cold water shower. Cooling fruit from about 35°C to about 18°C takes approximately 20 minutes.

Condensation, which can induce fruit splitting, favour decay and weaken paper-based packaging materials, is not a problem with pressure cooling. Cold air will progressively warm, increasing its water-holding capacity, as it passes through the package. In contrast, condensation may be a problem with room cooling, where transpired water vapour may condense on the faster-cooling outermost packaging and produce (Chapter 5).

Hydrocooling

In hydrocooling, water is the heat transfer medium. Thus, both produce and containers must be tolerant to wetting. Since water has a far greater heat capacity than air, hydrocooling is comparatively rapid provided that the water contacts most of the surface of the produce and is maintained close to the prescribed temperature, usually 0°C. In many hydrocooling systems, produce passes under cold showers on a continuous feed conveyor (Figure 4.5). Alternatively, cold water baths and/or batch process systems may be utilised. Hydrocooling may also help clean the produce. However, greater contamination of produce with spoilage microorganisms can occur if soil and debris are not removed (e.g. allowed to settle, or filtered) from the system, and the water renewed and/or disinfected. A further advantage of hydrocooling is that the commodity loses little weight during the process. When cooling is completed, produce must be moved to a cool room to prevent re-warming.

Icing

Before the advent of comparatively modern precooling techniques (e.g. pressure cooling), contact or package icing was used extensively for precooling produce and maintaining temperature during transit, particularly for the more perishable commodities such as leafy vegetables. The melting of ice to water (latent heat of fusion) absorbs 335 kJ (heat energy) per kg. Contact icing is now mainly employed as a supplement to other forms of precooling. In top icing, finely crushed ice or an ice slurry (liquid ice: 40% water, 60% ice, 0.1% salt) is sprayed onto the top of a load of produce, generally inside the road or rail transit vehicle. In some distribution and marketing systems, this practice is common, but it is largely unnecessary, expensive, and also undesirable due to the promotion of soft rots, to add crushed ice to precooled produce, such as broccoli and sweet corn packed in polystyrene containers. The alleged intention is to maintain freshness during transport and marketing. However, for the reasons mentioned above, the practice of package icing should be discouraged in favour of refrigerated handling and transport. Nevertheless, it is still used in transportation by air freight because of the difficulties of maintaining temperature control throughout the complex handling regime at airports.

Vacuum cooling

Vegetables such as lettuce, with a high surface to volume ratio, may be quickly and uniformly cooled by boiling off some of the constituent water at reduced pressure. The rate of cooling is at least as rapid as that achieved with hydrocooling. In principle, vacuum cooling is evaporative cooling, exploiting the latent heat of vaporisation of water. Produce is loaded into a sealed container, and the pressure is reduced to about 660 Pa (5 mm mercury) (Figure 4.6). At this sub-atmospheric pressure (cf. 670 mm mercury), water boils at 1°C. For every 5°C drop in temperature, approximately 1 per cent of the produce weight is lost as water vapour. This water loss may be minimised by spraying the produce with water either before enclosing it in the vacuum chamber or towards the end of the vacuum cooling operation (hydro-vacuum cooling). The rate of vacuum cooling is largely dependent on the surface to volume ratio of the produce, and on the ease with which the commodity loses water. Leafy vegetables are ideally suited to vacuum cooling. Other vegetables, such as asparagus, broccoli, Brussels sprout, mushroom and celery, can also be successfully vacuum cooled, as can cut flowers and foliage. Fruit with a low surface to volume ratio and/or a waxy cuticle loses water slowly and therefore does not benefit from vacuum cooling. Comparative cooling of several vegetables is shown in Table 4.2.

Figure 4.6 Hydro-vacuum cooler with capacity to cool four pallets of produce

(Supplied by Agsell Pty Ltd, Adelaide, South Australia.)

Table 4.2 Comparative cooling of vegetables from an initial temperature of 20°C under similar vacuum conditions

Produce	Final temperature (°C)
Lettuce, onion	2
Sweet corn	5
Broccoli	6
Asparagus, cabbage, celery, peas	7
Carrot	14
Potato, zucchini	18

SOURCE Adapted from American Society of Heating, Refrigerating and Air-conditioning Engineers (1986) *ASHRAE handbook of refrigeration systems and applications*. ASHRAE: Atlanta GA, USA.

Evaporative cooling

Evaporative cooling is a simple process in which dry air is cooled by blowing it across a wet surface. The evaporation of water (latent heat of vaporisation) absorbs 2260 kJ (heat energy) per kg. The technique is efficient only under conditions of low ambient RH (e.g. summer in Mediterranean-type climates; arid and semi-arid regions) and requires a good-quality water supply. It has the advantage of low energy cost. The commodity may be cooled either by the humidified cool air, or by misting with water then blowing dry air over the wet commodity. The extent to which air may be cooled by evaporation of water is a function of the water-holding capacity of the air, which in turn is a function of temperature and relative humidity (see Chapter 5). Evaporative cooling might be considered suitable for citrus, which does not require very low temperatures and is grown in dry environments such as Southern California, Israel and South Australia).

5 Water loss and Humidity

Fresh fruit, vegetables and ornamentals are comprised mostly of water (typically 80–90%). Accordingly, horticultural commodities might be regarded as water in attractive packages. It is costly to put water into these packages, and for produce sold by weight, water loss constitutes a direct financial loss. Just 5 per cent weight loss will cause many perishable commodities to wilt or shrivel. For some produce, this level of water loss can occur in just a few hours under warm and dry conditions. Even in the absence of wilting, water loss can equate to loss of quality, such as reduced crispness and other undesirable changes in colour, palatability, loss of nutritional quality (Figure 5.1) and impaired growth.

Wetting of produce can also result in substantial losses in some commodities. Free water encourages microbial decay and in some cases causes the produce to split (e.g. cherries). Additionally, free water can encourage undesirable growth, such as rooting and sprouting of produce such as potatoes and onions.

Terms

Postharvest water loss and associated aspects of plant and atmosphere water relations are easier to understand when one has a grasp of the basic terms used in the literature. Water loss is especially problematic, because most harvested horticultural produce is separated from the water supply that supported its growth on the plant.

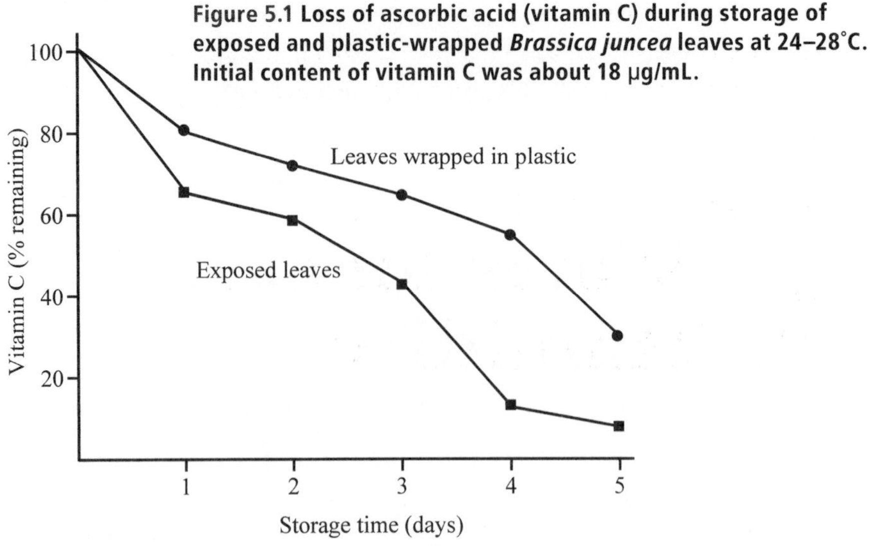

Figure 5.1 Loss of ascorbic acid (vitamin C) during storage of exposed and plastic-wrapped *Brassica juncea* leaves at 24–28°C. Initial content of vitamin C was about 18 μg/mL.

SOURCE Adapted from H. Lazan, Z.M. Ali and F. Nahar (1987) Water stress and quality decline during storage of tropical leafy vegetables. *Journal of Food Science* 52, pp. 1286–88.

Water potential

In plant tissue, turgor pressure (Ψ_{TP}) provides physical support and helps drive cell expansion. This is because water contained within the cells pushes against surrounding semi-permeable membranes and cell walls. The reason that water pushes against the membrane and wall is that more water is attracted to the cells than they can actually accommodate. Water is drawn into cells by their negative osmotic potential ($-\Psi_{OP}$), which in turn is due to dissolved organic (e.g. sugars) and inorganic (e.g. salt) solutes in the cell sap. Because positive turgor pressure (Ψ_{TP}) is smaller in magnitude than negative osmotic potential ($-\Psi_{OP}$), the two components balance to generate a tissue water potential that is negative ($-\Psi_{WP}$):

$$-\Psi_{WP} = \Psi_{TP} + (-\Psi_{OP})$$

For example, $-\Psi_{WP}$, Ψ_{TP} and $-\Psi_{OP}$ values for flesh tissue of avocado fruit were measured as –0.47, 0.19 and –0.65 MPa, respectively.

After harvest, water loss (transpiration) in the absence of water supply could soon dehydrate plant tissue. This is because the water potential of warm dry air is lower (more negative) than that of the plant tissue. Loss of water leads to wilting when turgor falls to zero (Figure 5.2). However, wilting may not be immediately obvious in all tissues, such as those made up of thick-walled cells. Moreover, water loss from harvested produce is slowed by structural barriers (e.g. waxy cuticle) and physiological controls (e.g. stomatal closure), which confer significant resistance to water loss in most crops.

Vapour pressure (VP)

Humidity is the general term referring to the presence of water vapour in air. If pure water is placed in an enclosure containing dry air, water molecules will enter the vapour phase until the air becomes saturated with water vapour. The pressure that these molecules exert on the container walls is the vapour pressure (VP). The amount of water vapour in atmospheric air

Figure 5.2 A Hofler diagram derived from analysis of the Ψ_{WP} isotherm and showing the relation between relative water content and water potential (Ψ_{WP}), osmotic potential (Ψ_{OP}) and turgor potential (Ψ_{TP})

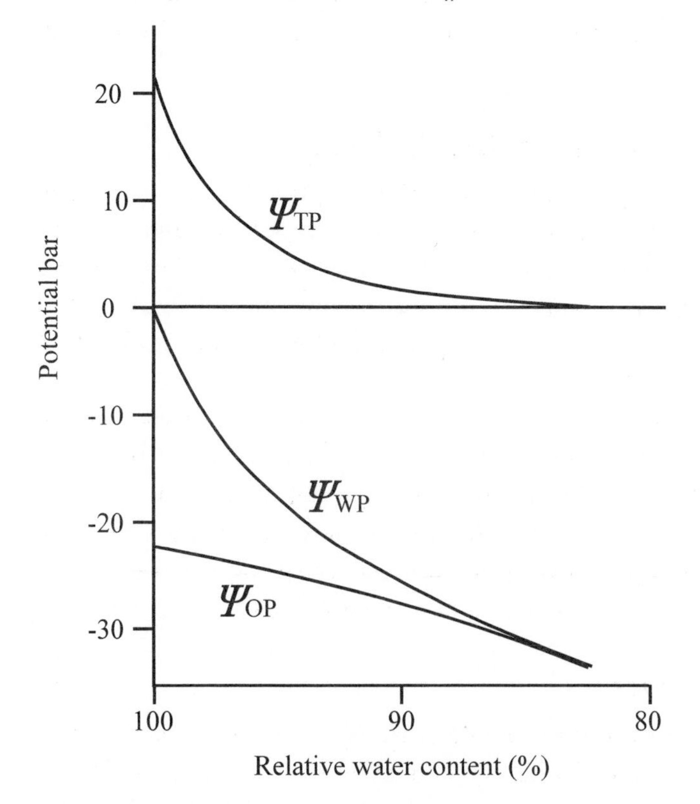

SOURCE Adapted from M.T. Tyree and P.G. Jarvis (1982) Water in tissues and cells, in O.L. Lange, P.S. Nobel, C.B. Osmond and H. Ziegler, *Physiological plant ecology,* vol. II, *Water relations and carbon assimilation*. Springer-Verlag: Berlin, Germany.

can vary from zero to a maximum that is dependent on temperature and pressure. For example, saturated air at 30°C contains approximately 30.4 g water/m^3 of air and the VP is about 4.25 kPa.

Relative humidity (RH)

RH is the best known term for expressing the moisture content of air. RH is defined as the ratio of the water vapour pressure in the air to the saturation vapour pressure possible at the same temperature, and is expressed as a percentage (%). The RH of saturated air is 100%.

The water potential of air is related to RH by the following logarithmic relationship:

$$-\Psi_{WV} = (RT/V_W) \times \ln(RH\%/100)$$

where Ψ_{WV} is water potential of water vapour, V_W is the molar volume of water (e.g. 18.048 x 10^{-6} m^3/mol. at 20°C), R is the gas constant, and T is the absolute temperature. The water potential of air is usually low relative to that of harvested horticultural produce (e.g. –14.2 MPa at 90% RH and –93.6 MPa at 50% RH; 20°C).

Equilibrium relative humidity (ERH)

When water-containing plant tissue material is placed in an enclosure filled with air, the water content of the air increases or decreases until equilibrium relative humidity (ERH) is reached. At this point, the number of water molecules entering and leaving the vapour phase is equal. ERH is a property of the moisture content of the individual plant tissue. Pure water has an ERH of 100%.

Measures of ERH are of particular interest to microbiologists, because small reductions in water activity (a_W; where a_W = ERH/100) and water potential equivalent to a reduction in ERH to about 95% can inhibit the growth of most bacteria and fungi in culture. However, growth of some pathogenic fungi is not prevented until water activity and water potential are reduced to less than the equivalent of about 85% ERH.

Psychrometric chart

Psychrometric charts graphically relate various properties (e.g. wet bulb and dry bulb temperatures, water vapour pressure) of moist air to one another (Figure 5.3). The scale along the bottom axis of Figure 5.3 indicates dry bulb temperatures as given by a wet-and-dry bulb hygrometer. At all temperatures, dry air has no water and, therefore, a water vapour pressure

Figure 5.3 Simplified psychrometric chart

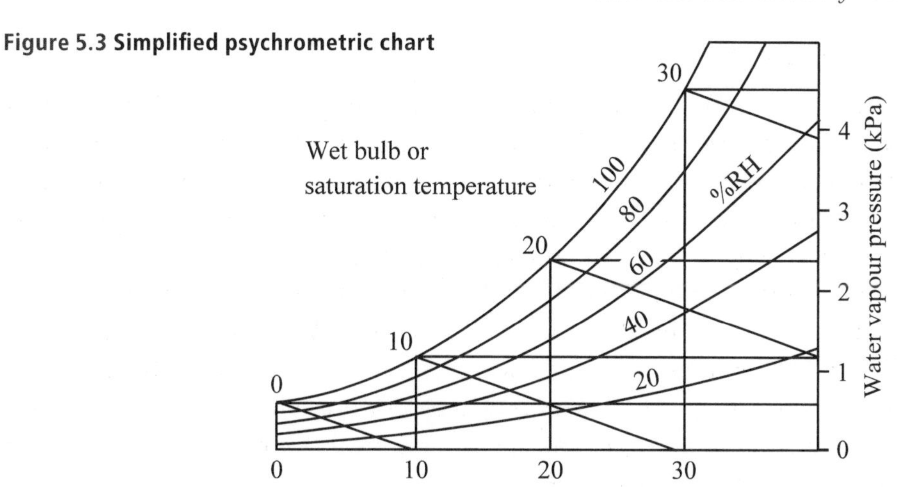

of zero. The curved line at the top of Figure 5.3 illustrates the relationship between VP and temperature in saturated air, and is the line of 100% RH. Other curved lines can be drawn representing constant RH over a range of temperatures. The curvature of these lines shows that the VP of water increases rapidly with temperature. For example, at 30°C in saturated air (100% RH) the VP of water is 4.3 kPa; at 20°C the VP is only 2.4 kPa; and at 10°C it is merely 1.3 kPa.

Vapour pressure deficit (VPD)

The difference between the VP of the produce, which is a function of temperature and ERH, and that of the surrounding air, which is a function of temperature and RH, is called the vapour pressure deficit (VPD). VPD calculations can be used to compare relative rates of water loss under different sets of conditions. Consider, for example, produce at 20°C and 97% ERH and humidified air at 0°C and 100% RH (condition 1). The VP of the produce will be about 2.4 kPa, whereas that of the air will be about 0.6 kPa (Figure 5.3). Thus, the VPD will be 1.8 kPa (i.e. 2.4 kPa [produce] – 0.6 kPa [air]), and water will be lost from the produce.

Consider now the relative rate of water loss for produce already cool at 5°C (97% ERH) and being placed into the same airstream (0°C, 100% RH) (condition 2). The VP of this produce will be 0.9 kPa and the VP of the airstream will still be 0.6 kPa. Thus, the VPD for this second set of conditions will be 0.3 kPa (i.e. 0.9 kPa – 0.6 kPa). Accordingly, initial water loss under condition 1 is likely to occur at about 6 times the rate of water loss under condition 2 (i.e. 1.8 kPa/0.3 kPa = 6).

Condensation

Dewpoint is another important physical property of moist air that is evident in psychrometric charts. When moist air is cooled, a temperature is reached at which the VP of the air reaches the maximum for that temperature. Thereafter, with further cooling, water will condense. For example, fog is formed (water condenses) on a cooled surface. The temperature at which condensation occurs is the dewpoint temperature. The horizontal lines in Figure 5.3 can be used to find dewpoint temperatures. Thus, air of 80% RH at 30°C becomes saturated when cooled to about 26°C. The dewpoint temperature is equal to the dry bulb temperature at the point of intersection with the saturation curve. The lines that slope upwards from right to left in Figure 5.3 indicate constant wet bulb temperatures.

If a fibreboard carton of warm, transpiring produce is placed in a cold room, the carton will cool relatively quickly and water will condense onto it because condensation forms on cold surfaces. The damp carton may collapse under the weight of other cartons, resulting in physical damage to the produce, as the strength of fibreboard is substantially reduced when it is moistened. In an alternative scenario, condensation will form on cooled produce in warm moist air (e.g. warming carton). This condensation (free water) can promote rots (i.e. spore germination), accelerate warming of the produce (i.e. latent heat of vaporisation release, high thermal conductivity),

Table 5.1 Transpiration coefficients for selected horticultural produce

Product	Transpiration coefficient (mg kg.sec^{-1} MPa^{-1})	Range (from literature)
Apple	42	16–100
Brussels sprout	6150	3250–9770
Cabbage	223	40–667
Carrot	1207	106–3250
Grapefruit	81	29–167
Lettuce	7400	680–8750
Onion	60	13–123
Peach	572	142–2089
Potato	25	15–40
Tomato	140	71–365

SOURCE S. Ben-Yehoshua (1987) Transpiration, water stress, and gas exchange. In J. Weichmann (ed.) *Postharvest physiology of vegetables*. Marcel Dekker: New York NY, USA.

induce splitting (i.e. increase turgor) and interfere with respiratory and other (e.g. ethylene) gas exchange (the diffusion of gas in water is 10 000 times slower than in air). Under conditions of low storage temperature (e.g. 0°C) and high RH (e.g. 95%), extremely small fluctuations in temperature (<0.5°C) can result in condensation on cooling surfaces.

Factors affecting water loss

Horticultural produce varies widely in the physicochemical characteristics that determine inherent rates of postharvest water loss. Some examples of relative rates of water loss are presented in Table 5.1.

Produce characteristics

Surface area/volume ratio

A major determining factor in the rate of water loss from produce is the surface area to volume ratio. There is proportionally greater loss by transpiration from produce with a high surface area to unit volume ratio. Individual edible leaves have surface area (SA) to volume (V) ratios of around 50–100 $cm^2.cm^{-3}$, whereas tubers have SA:V ratios of about 0.5–1.5 $cm^2.cm^{-3}$. Thus, other factors being equal, a leafy vegetable (e.g. spinach) or a cut flower (e.g. rose) will lose water much faster than a fruit (e.g. apple). Similarly, a small fruit, root or tuber will lose water faster than a relatively large one.

Plant surfaces

The structure and composition of plant surfaces and underlying tissues has a marked effect on the rate of water loss. Many types of produce have a waxy cuticle on the surface. The cuticle is typically resistant to the passage of water or water vapour and plays an essential role in restricting water loss by evaporation and maintaining high water content within the tissue. It has been estimated that the cuticle reduces the rate of evaporation from living plant cells about 25-fold.

The structure of the wax coating may be more important than its absolute thickness. Waxy coatings that consist of well-ordered overlapping platelets provide greater resistance to the permeation of water vapour than coatings without a tertiary structure. In the former case, water vapour must follow a more complex path as it escapes to the atmosphere. An unstirred boundary layer of air over the surface of produce can be thickened by the presence of hairs (trichomes). Trichomes vary widely in shape (e.g. club, branched) and structure (e.g. single or multi-celled). They derive from the epidermis, which underlies the waxy cuticle. Epidermal cells are compactly structured, with minimal space in the walls between adjacent cells.

The bulk movement of water vapour and other gases (oxygen, carbon dioxide) into and out of leaves is controlled by small pores called stomata (singular: stoma; Figure 5.4). These pores are located in the epidermis. Sub-stomatal cavities lie below the stomata and connect with the intercellular airspace network. Stomata in harvested leafy produce normally close when the surrounding pair of guard cells lose turgor in response to a small amount of water loss. They also close in response to darkness. However, under some conditions stomata may remain open, such as during rapid cooling of chilling-sensitive tissues.

Many fruit (e.g. avocado) and storage organs (e.g. sweet potato) have lenticels, and not stomata, on their surface. Lenticels are sunken openings that may originate from stomata. However, they are larger and generally contain tightly packed and suberised hypodermal cells. Suberin, like cutin in the cuticle, is a hydrophobic compound that serves to minimise water loss. Otherwise, there is no mechanism for closing lenticels. Lenticels are often blocked, particularly in mature fruit, with wax and debris. In this

Figure 5.4 Cross-section of a leaf showing the surface (cuticle), stoma, internal structures and the network of intercellular spaces

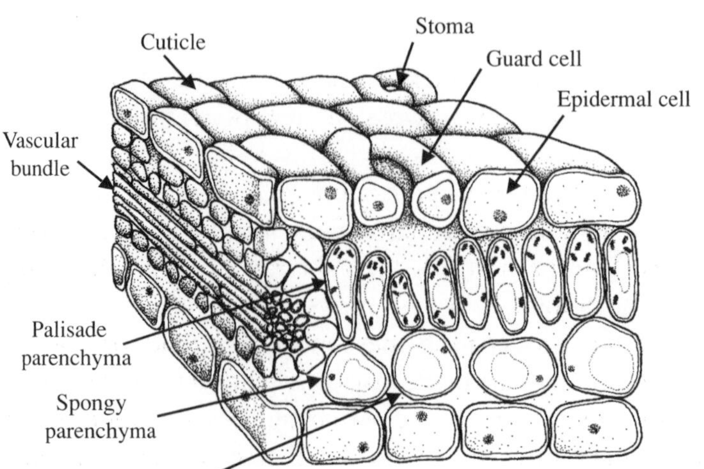

case, loss of vapour water, and also respiratory gas exchange, may only take place by diffusion through the cuticle. In some harvested fruit (e.g. tomato), a large stem scar can also be an important point of water loss.

The surface of some tubers and roots is a periderm made of several layers of suberised (cork) cells. Periderm is produced from a layer of cambial cells. It is important in limiting water loss when underground organs (e.g. potato tubers) are exposed to relatively low above-ground humidity after harvest. Curing is a postharvest process to encourage periderm formation. For example, harvested sweet potato can be cured under high humidity (90–95% RH) and elevated temperature (29°C) conditions for about 1 week.

Mechanical injury

Mechanical damage can accelerate the rate of water loss from harvested produce. Bruising and abrasion damage the surface structure, allowing greater evaporation. Cuts both break the protective surface layer and directly expose underlying tissues to the atmosphere. Damage to surface tissue can also occur as a consequence of attack by pests (insects, rodents) or diseases, and will result in increased rates of water loss. If damage occurs early in the growth and development of fresh produce, the organ may seal the affected area with a layer of corky callus cells. However, the capacity for wound healing generally diminishes as plant organs mature. Thus, damage that occurs during harvesting or postharvest operations generally remains unprotected. Nonetheless, some types of mature produce (e.g. tubers and roots) retain the ability to seal wounds. Curing at suitable temperatures and humidities encourages wound-healing (see above). If concentrated solutes (e.g. sugars) are released upon wounding, they may osmotically attract water vapour. A droplet will then grow in volume under conditions of high in-package RH and continued release of solutes from the wound. In due course, rots will develop due to microbial growth in the droplets and wound.

Management of water loss

Covering cracks on harvested produce with wraps or surface coatings (e.g. waxed apples) or harvesting fruit at a more advanced stage of maturity (e.g. marrows versus zucchini) are ways of modifying and utilising, respectively, tissue structure to limit water loss. However, there is only limited scope for reducing rates of water loss in these ways. The most significant methods of reducing water loss from produce primarily entail lowering the capacity of surrounding air to hold additional water. This can be achieved by lowering temperature and/or raising RH, bringing about a reduction in the VPD between the product and air (Figure 5.5).

Relative humidity (RH)

Increasing the RH of the air reduces the VPD and, thus, the amount of water lost from produce before the surrounding air is saturated. However, very high RH (>95%) can encourage the growth of bacteria or fungi. Promotion of rots at very high humidity may outweigh the potential benefits of reduced water loss for certain produce. Most fungi cease to grow when the RH is reduced to about 90%, and only a few can grow at 85%. Under drier conditions spores cannot germinate and, even if there is enough free moisture in a wound to permit germination, a dry atmosphere may dehydrate exposed tissue fast enough to prevent infection and development of a rot. Experience has shown that RH of 90% is usually the best compromise condition for fruit storage. However, 98–100% RH is the best condition for leafy vegetables, some root vegetables, and for cut flowers and foliage, because these have relatively high coefficients of

Figure 5.5 Effect of vapour pressure difference (deficit) on weight loss by the apple cultivars Golden Delicious and Jonathan

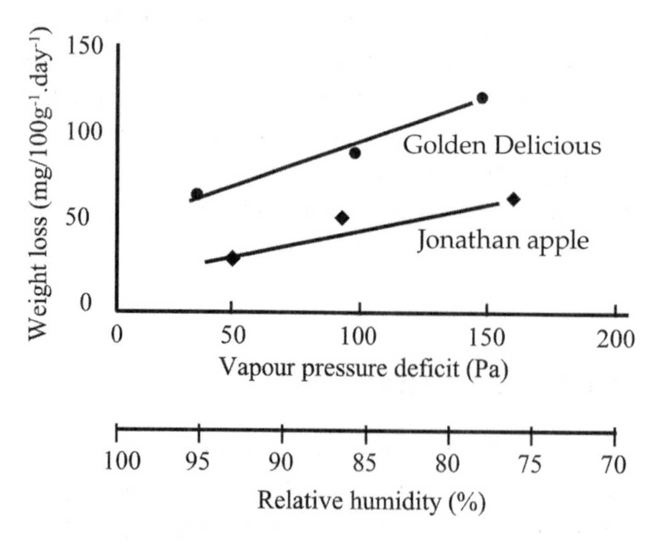

SOURCE Adapted from A.W. Wells (1962) *Effects of storage temperature and humidity on loss of weight by fruit*. US Department of Agriculture: Washington DC, USA, Marketing Research Report no. 539.

transpiration. Fungicides may be used to minimise or overcome a problem of fungal growth at high RH. However, consumer resistance to the use of such chemicals should be considered. Storing potato continuously at high humidity has the added advantage that it reduces the likelihood of pressure bruises developing, compared to storage at low humidity. However, commodities such as potato and onion are also more likely to sprout at high RH. In contrast to most harvested horticultural produce, onion and mature cucurbits (e.g. pumpkins) require a low in-storage RH of 65–70%, which prevents excessive rotting.

It is relatively simple to increase the RH of air. Spraying water as a fine mist, introducing steam, and increasing the temperature of the refrigeration coils are practical means. Addition of water vapour to a cold storage chamber can be controlled automatically with a humidistat. However, adding free water to a refrigerated system may result in condensation on the cold surfaces of produce, walls and cartons and also pooling on floors. Moreover, the increased frosting-up of refrigeration coils will necessitate longer and more frequent defrost cycles. For produce in cold rooms, the best way to maintain high RH (95%) is to maintain a small temperature difference between the refrigeration coils and the produce. This is done by using large coils with a high surface area for heat exchange (Chapter 7). Another approach is to use an air-wash refrigeration system (Chapter 7).

In the context of controlled ripening, fruit will have a better appearance (less shrivelling) and better internal quality if kept at RH greater than 90%. The need to control RH in banana ripening rooms to prevent splitting and shrivelling is well recognised (Chapter 11). In contrast, storage at very high humidity of apple varieties susceptible to internal or low temperature breakdown only enhances development of the disorder due to a low level of water loss. Storage at a lower humidity enhances water loss, which has the effect of decreasing the incidence of this disorder.

Air movement

Air movement over produce significantly influences the rate of moisture loss. Therefore, while air movement is required to remove heat from produce, its effect on moisture loss must also be considered. There is always a thin layer of unstirred air (the boundary layer) at the surface of produce. In this layer, the water VP is approximately in equilibrium with that of the produce itself. Increasing the rate of air movement reduces the boundary layer thickness, increasing the VPD near the surface and, thus, the rate of moisture loss. Restricting air movement around produce in a cool store can effectively reduce water loss.

Rapid air movement is required initially for fast cooling of warm produce. However, reduced air movement can subsequently be achieved by running fans on the evaporator at a lower speed or by reducing the periods of time for which they operate (Chapter 7). If storage is in open rooms with natural ventilation, modifications can be adopted to restrict air flow. Regulation of air movement requires compromise. Sufficient air movement is needed to prevent significant temperature gradients forming within the storage chamber, but it should be controlled at rates that minimise water loss from produce. Maintaining a high RH by eliminating both VPD and air movement in a cool store system can be achieved by circulating cold air around a sealed plastic tent that contains the produce. On a smaller scale, the same condition is achieved with plastic pallet shrouds and carton liners.

Air pressure

Reduced air pressure, as may be experienced by horticultural produce during vacuum cooling and air freight under partial pressurisation conditions, will increase the rate of water loss. Thus, the time that produce spends under reduced pressure conditions must be minimised, and steps (e.g. misting with water, provision of moisture barrier packaging) taken to inhibit water loss.

Packaging

Water loss can be effectively reduced by placing a barrier around produce, which also prevents surface air movement. Simple methods are to pack produce into bags or boxes and to cover stacks of produce with tarpaulins. Close packing of produce itself restricts the passage of air around individual items and, thereby, reduces water loss. Thus, even placing produce in mesh bags can be beneficial because the closer packing leads to thick unstirred boundary layers around inner produce items.

The degree to which water loss is reduced by packaging depends on the permeability of the package to water vapour transfer, as well as on the closeness of containment. Most common packaging materials are permeable to water vapour to some extent. However, plastic (e.g. polyethylene) films are excellent vapour barriers. Their rate of water vapour transfer is extremely low compared to that of, say, paper. Nevertheless, even paper bags and fibreboard packages will substantially reduce water loss compared with loose produce. It should be borne in mind that packaging also reduces the rate of cooling, by restricting air movement around individual items.

The use of very thin plastic wrap and heat-shrink films for individual packaging can significantly increase the storage life of many fruits by greatly reducing their water loss. Such films offer storage in a water-vapour-saturated atmosphere, without the problem of condensation. Theoretically, condensation cannot occur because the film is very thin (e.g. 10–50 μm) and the produce and the film are in intimate contact and, therefore, at exactly the same temperature. However, there are often areas where the produce and the film are not in contact (e.g. stem scars). Condensation can occur in these locations and, in turn, promote fungal development. Consequently, fruit (e.g. citrus, melons) is generally treated with approved fungicides before it is packaged in films. In this system, the spread of decay from any fruit that might become infected is effectively, and conveniently, prevented by the film itself. There is evidence that healing of skin injuries inflicted during harvesting and packing-house operations to citrus fruit is promoted by the saturated atmosphere created within heat-shrink films. If the film is not fully sealed due to incomplete coverage, inadvertent holes or deliberate macro- or micro-perforation, the wrapping has little effect on the fruit's internal respiratory gas composition atmosphere. Accordingly, there is little risk of anaerobiosis and associated quality defects, such as off-flavours and discolouration. Because heat-shrink film is in direct contact with the surface of individual fruit, there is virtually no effect on the rate of heat exchange.

It is important to consider the ability of many packaging materials to absorb moisture vapour and water. Paper derivatives, jute bags and other natural fibres can absorb considerable amounts of moisture before becoming visibly damp. At the time of packing, there is often a significant VPD between the produce and the package. Thus, water evaporated from the produce may be absorbed by the packaging material. In the cool storage of apple and pear fruit, it was found that a dry wooden box weighing 4 kg would absorb about 500 g of water at 0°C. Ideally, packages should be equilibrated at high humidity before use, but this is impractical commercially. An alternative procedure is to waterproof moisture-absorbent packaging materials like fibreboard with wax or resin coatings, or to alternatively use a non-absorbent packaging material like polystyrene.

The ability of certain packaging and other materials to absorb and desorb moisture can be exploited to achieve moisture management within packages. For instance, inorganic salts placed in sachets permeable to water vapour, but not to free water, can help control in-package RH. Such moisture sinks may be used to lower RH and avoid condensation within a package. In turn, the risk of disorders such as fruit splitting or decay is reduced. Conversely, moisture sinks with sufficient sorption capacity can work as reservoirs or moisture stores. Such moisture stores can supply water vapour to dehydrating produce within a package. As a practical example, dry paper packaging can be used to inhibit the growth and development of grey mould on roses. On the other hand, wrapping in moistened paper can reduce bent neck, a disorder associated with excessive water loss from the rose flower peduncle.

Ornamentals

The water relations of ornamentals warrant special attention. Unlike fruit and most vegetables, ornamentals are often supplied with water for part (e.g. cut flowers and foliage) or all (e.g. pot plants) of the post-production period. It is important to keep the growing medium of most pot plants moist to avoid wilting and water stress. To this end, high water-holding capacity ingredients (e.g. vermiculite, peat, hydrophilic synthetic polymers) are usually included in the rooting medium. Nonetheless, during long transits and periods in confined spaces, reduction in RH by ventilation or some other means is essential in order to minimise fungal and bacterial diseases.

Termination of the display or vase life of fresh cut flowers and foliage is often associated with wilting. Wilting, as for pot plants, generally occurs when transpiration demand exceeds water supply. Thus, wilting typically occurs because of an increased resistance to water flow up the stem. Adverse display conditions, such as direct sunlight and dry air-conditioned air, will increase VPD and thus exacerbate problems associated with resistance to water uptake and flow. Resulting from an increase in stem resistance to water flow, the column of water above the limiting resistance will come under increasing tension. In turn, cavitation may occur, with the consequence that resultant air bubbles (emboli) further restrict stem flow.

Resistance to stem water flow can be caused by physical, physiological or biological agents. Physical agents include air bubbles entrapped in xylem cells at the cut end of the stem and plugging with inorganic (e.g. clay) and organic (e.g. dead bacteria, cell wall fragments) particles. Ingress of air into cut xylem endings can be avoided by recutting the stem ends under water. Physiological plugging involves active metabolism in wound healing processes that attempt to seal off the cut stem end. Formation of tyloses in the lumen of xylem vessels is an example of physiological plugging. Biological plugging is associated with plugging by live (and dead) microbes. Some live microbes may attack the structure of water-conducting and adjacent plant tissues. Factors determining the detrimental effects of different microbes include their size, formation of extracellular slime matrices, and secretion of enzymes (e.g. pectinases). Nevertheless, cut flower and foliage stems with prolific microbial growth on and around the cut end in the vase solution may still retain turgor. In such instances, plugging might be incomplete, stomata could be closed and water may flow via alternative routes, such as unsealed lenticels on the stem and cell-to-cell or cell-wall pathways. Cut flower and foliage species with characteristically high rates of water use (e.g. roses) appear to be relatively more susceptible to wilting.

A wide variety of chemicals are used in vase solutions to help maintain a positive water balance. Acidifiers such as citric acid are used to enhance water flow, although the mechanism of action is unknown, and to reduce pH below that which supports the optimal growth of most microbes. Biocides, such as chlorine and quaternary ammonium compounds, are used to kill microbes. Chemicals that suppress ethylene synthesis (e.g. aminoethoxyacetic acid [AOA]) or action (e.g. silver thiosulphate [STS]) may possibly inhibit physiological plugging. Abscisic acid has been used to induce stomatal closure and, thereby, increase resistance to water loss from leaves. Abscisic acid and film-forming anti-transpirants may be applied as a dip or spray. Compatible inorganic (e.g. potassium chloride) and organic (e.g. sucrose) solutes can be provided in vase solutions to lower the osmotic potential of cells, thereby aiding osmotic adjustment and turgor maintenance. Surfactants that decrease resistance to water flow in the xylem, and perhaps promote dissolution of air bubbles, can also be used in vase solutions to help preserve the water balance of cut flowers and foliage. With chemicals used to maintain water balance, it is important to determine their effective concentrations alone and in combination with other ingredients for specific fresh-cut flower and foliage lines, and to avoid phytotoxicity.

Roses are the most economically important of all cut flowers. High tissue water status must be maintained during their transport and handling. The elongating peduncle is poorly lignified and turgor maintenance is critical in supporting the rose bud. If turgor is lost, then cells in the peduncle may be damaged and blacken. Thereafter, the peduncle cannot support the rose bud, giving rise to the bent neck disorder. On the other hand, it is practice to slightly wilt kangaroo paw inflorescences prior to their packaging and transport. As a result of increased flexibility, fewer flowering cymes are broken off the inflorescence during handling.

Water relations are also an issue with dead, preserved ornamental material. Humectants, such as glycerol, are impregnated into the plant material either by immersing it in vats of solution or by uptake through transpiration. These humectants attract water vapour from the atmosphere, maintaining otherwise dry plant material in a supple, plasticised condition. However, if preserved material is treated with too much glycerine, it will exude droplets of glycerol solution at high RH. Apart from being unsightly and sticky, this sweating supports the growth of moulds.

6 Storage atmosphere

The composition of gases in the storage atmosphere can affect the storage life of horticultural produce. Alteration in the concentrations of the respiratory gases, oxygen and carbon dioxide, may extend storage life. This is generally used as an adjunct to low temperature storage, but for some commodities modification of the storage atmosphere can usefully substitute for refrigeration. Many volatile compounds generated by produce and from other sources may accumulate in the storage atmosphere. Ethylene is the most important of these compounds and if it accumulates above certain critical levels it reduces storage life, so controlling ethylene is important.

The terms controlled atmosphere (CA) storage and modified atmosphere (MA) storage are frequently used. These terms imply the addition or removal of gases, resulting in an atmospheric composition different from that of normal air. The levels of carbon dioxide, oxygen, nitrogen and ethylene in the atmosphere may be manipulated. CA storage generally refers to decreased oxygen and increased carbon dioxide concentrations and implies precise control of these gases. In MA storage, the composition of the storage atmosphere is not closely controlled (for example, in plastic film packages where the change in the composition of the atmosphere occurs intentionally or unintentionally), although MA storage is often used as a generic descriptor of atmosphere change. A more recent term is MAP, modified atmosphere packaging, which relates to packages and film box liners with specific properties that offer some measure of control over the composition of the atmosphere around produce.

Carbon dioxide and oxygen

The general equation for produce respiration:

glucose + oxygen —> carbon dioxide + water

suggests that respiration could be slowed by limiting the oxygen or by raising the carbon dioxide concentration in the storage atmosphere. The principle appears to have been applied in ancient times, even if unwittingly. The earliest use of MA storage may possibly be attributed to the Chinese. Ancient writings report that litchi was transported from southern China to northern China in sealed clay pots to which fresh leaves and grass were added. It may be surmised that during the two-week journey, respiration of the fruit, leaves, and grass generated a high carbon dioxide–low oxygen atmosphere in the pots, which retarded ripening of the litchis. Other examples of primitive MA storage include the burying of apples in the ground and the carriage of fruit in the unventilated holds of ships. The first reported scientific observations of the effects of atmospheres on fruit ripening were made in 1819–20 by Jacques Berard in France, but it was not until the work of Kidd and West at the Low Temperature Research Station at Cambridge, UK, in the 1920s and 1930s, that a sound basis for the controlled atmosphere storage of produce was established.

The effects of modified atmospheres have since been extensively tested on a wide range of fruit and vegetables, but the responses have varied considerably. Despite extensive research, major commercial application of controlled atmosphere has been confined to some apple and pear cultivars, but modified atmospheres have been applied successfully during transport to a range of produce. For example, high carbon dioxide levels have been used primarily as a fungistat during the transport of strawberries, and improved out-turns of lettuce have been achieved by flushing rail trucks or containers with nitrogen and up to 8 per cent carbon monoxide. These techniques are of considerable relevance for minimally processed produce, especially lettuce, due to the short shelf-life and susceptibility to browning (see Chapter 11).

As with fruits and vegetables in general, MA storage of ornamentals has not been widely implemented. Nonetheless, atmosphere recommendations have been established for a number of important cut flower crops (Table 6.1). In addition to extended longevity at low temperatures, storage under MA conditions, particularly with high carbon dioxide, can limit the development of pathogens. The lack of commercial use arises from considerable response inconsistencies within the same floral species across different seasons and different varieties. In addition, the regular production schedules of major species and limited quantities of minor species create little need to store many cut flowers for significant periods.

Table 6.1 Optimal controlled atmosphere conditions for certain flowers

Species	% CO_2	% O_2	Temp (°C)	Storage period
Freesia	10	21	1	3 weeks
Carnation	5	2	0	4 weeks
Lily	10–20	21	1	3 weeks
Mimosa	0	8	7	10 days
Rose	5–10	1–3	0	3–4 weeks

SOURCE D.M. Goszczynska and R.M. Rudnicki (1988) Storage of cut flowers. *Horticultural Reviews* 10, pp. 35–62.

Factors that have influenced the adoption of modified atmospheres for different commodities include:

- Inherent storage life in air. If the produce can be stored in a satisfactory condition in air for the total marketing period desired, then there is no need to resort to atmosphere modification to prolong storage life.
- Existence and magnitude of a favourable response to modified atmospheres. There must be a distinct beneficial effect. Not all produce responds favourably to atmosphere regulation and some produce is little affected.
- Substantial atmosphere tolerance. The beneficial effects of atmosphere modification, especially in non-controlled atmosphere storage, need to be sustained over a relatively wide gas concentration range. A small tolerance range can result in variable quality out-turns in commercial usage due to insufficient or excessive gas concentrations.
- Seasonal availability. Use of atmospheres can be advantageous where produce is harvested over a relatively short period in the year. Maximum storage life of such produce is often desirable to extend the marketing period.
- Value of the commodity in relation to the cost of atmosphere modification. There needs to be a distinct financial gain from the use of atmosphere control.
- Availability of substitute commodities. While produce may be stored satisfactorily in modified atmospheres, it may be more economical to import produce from another region or country that has a different harvest period.

The use of modified atmospheres in portable packages should have considerable commercial appeal, but the lack of widespread application is largely due to problems in ensuring that the desired atmosphere is maintained throughout the postharvest chain under a diverse range of handling operations and external environmental conditions. A variable MA will often result in variable produce quality in the market. The availability of a range of new food-grade polymeric films with differing permeability to the atmospheric gases in recent years has revived interest in packaging produce in sealed bags. The use of such films for refrigerated produce offers a cheaper alternative to using large containers equipped to provide modified or controlled atmosphere conditions. The development and properties of these plastic films will be covered in greater detail in Chapter 12.

Metabolic effects

Increases in carbon dioxide and decreases in oxygen concentrations exert largely independent effects on respiration and other metabolic reactions. Generally, the oxygen concentration must be reduced to less than 10 per cent (by volume) before any retardation of respiration is achieved. For apples stored at 5°C, the oxygen level must be reduced to about 2.5 per cent to achieve a 50 per cent reduction in respiration rate. Care must also be taken to ensure that sufficient oxygen is retained in the atmosphere so that anaerobic respiration, with its associated development of off-flavours, is not initiated.

The reduction in oxygen concentration that is necessary to retard respiration depends on the storage temperature. As the temperature is lowered the required concentration of oxygen is also reduced. The critical oxygen level beyond which anaerobic respiration occurs is determined mainly by the rate of respiration; therefore it is greater at higher temperatures. Tolerance to low oxygen levels varies considerably among different commodities. The critical level of oxygen may vary with the time of exposure, with lower levels being tolerated for shorter periods. It may also be affected by the level of carbon dioxide, since lower levels of oxygen often seem to be better tolerated when carbon dioxide is absent or at a low level.

The addition of only a few per cent of carbon dioxide to the storage atmosphere can have a marked effect on respiration. However, if carbon dioxide levels are too high, effects similar to those caused by anaerobiosis (lack

Figure 6.1 Relative tolerance of fruit and vegetables to elevated carbon dioxide and reduced oxygen concentrations at recommended storage temperatures. Normal atmospheric air comprises 0.036% carbon dioxide, 21% oxygen and about 79% nitrogen.

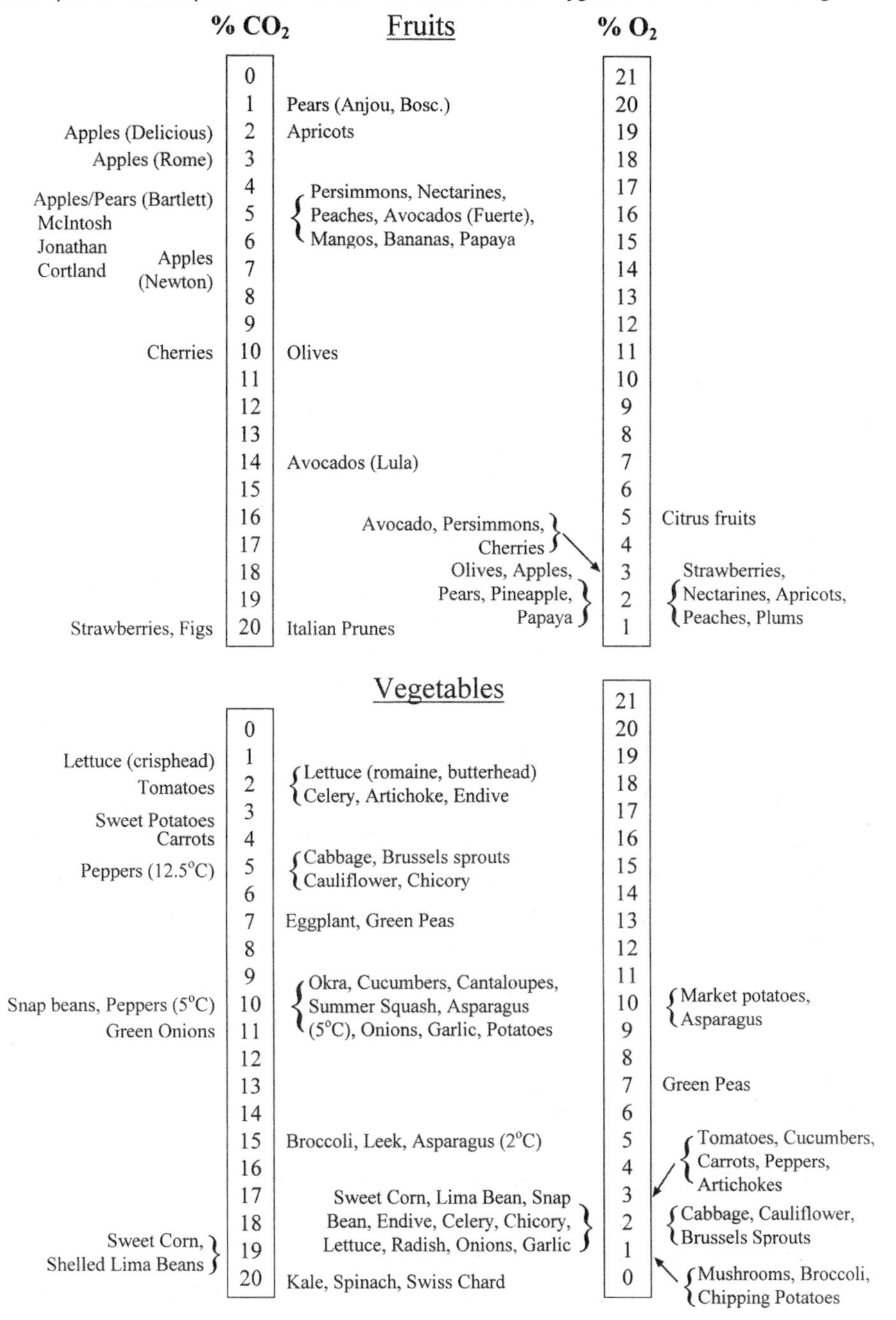

SOURCE A.A. Kader and L.L. Morris (1977) Relative tolerance of fruits and vegetables to elevated CO_2 and reduced O_2 levels. In D.H. Dewey (ed.) *Controlled atmospheres for storage and transport of horticultural crops*. Michigan State University: East Lansing MI, USA, pp. 260–65.

of oxygen) can be initiated. Responses to increased carbon dioxide levels vary even more widely than responses to reduced oxygen: cherries and strawberries will withstand, and even benefit from, exposure to 30 per cent carbon dioxide for short periods. In contrast, some apple cultivars are injured by 2 per cent carbon dioxide in storage, and many vegetables appear to respond best to low oxygen when carbon dioxide is kept low or is absent (Figure 6.1, p.87).

Many of the beneficial results of MA storage cannot simply be attributed to a reduction in respiration. For example, under ideal experimental conditions at 20°C a 12-fold increase in the storage life of green banana can be achieved by ventilating the fruits with an atmosphere comprising 5 per cent carbon dioxide, 3 per cent oxygen and 92 per cent nitrogen, in the absence of ethylene, but respiration measured in terms of oxygen uptake is reduced to only one-quarter of the rate in air. The greatly increased storage life is attributed to a reduced rate of natural ethylene production by the bananas and a reduced sensitivity of the fruit to ethylene.

In green vegetables, improved retention of green colour in low oxygen atmospheres is due mainly to a lowered rate of chlorophyll destruction. An interesting and contrasting effect has been noted in potato: greening due to light exposure can be prevented for several days by maintaining tubers in an atmosphere containing about 15 per cent carbon dioxide.

Modified atmospheres, particularly those high in carbon dioxide, inhibit breakdown of pectic substances so that firmer texture is retained for a longer period. Retention of flavour may also be improved. The responses of various commodities to modified atmospheres can, however, be mixed. For example, increased carbon dioxide helps retain organic acids in tomato but accelerates loss of acids in asparagus. Maturity at harvest is more critical for modified atmosphere storage than for ordinary air storage. Because of the widely varying responses of different commodities, and among cultivars, to alterations in oxygen and carbon dioxide concentrations, ideal combinations need to be determined experimentally for each commodity.

Effect on microbial growth

The activity of several decay organisms can be reduced by atmospheres containing 10 per cent carbon dioxide or more, provided that the commodity is not injured by such high carbon dioxide levels. Since strawberries can tolerate high carbon dioxide, transporting them under a modified atmosphere has been found to significantly reduce rotting and improve market life and quality.

Many commodities cannot tolerate high carbon dioxide levels, so in practice atmosphere control cannot always be relied on to reduce rotting. Nevertheless, use of a modified atmosphere may reduce rotting by retarding ripening and senescence, since the natural resistance of the produce host to pathogens decreases as it ripens or ages. In contrast, some fruits, such as banana and mango, respond well to atmosphere control but eventually lose their resistance to the latent anthracnose disease, which then becomes the factor limiting storage life. MA storage does not necessarily retard ageing and the loss of resistance to decay organisms at the same rates.

Methods for modifying carbon dioxide and oxygen concentrations

Controlled atmosphere (CA) storage

The earliest commercial applications of CA storage relied on the produce itself to generate the atmosphere, so that carbon dioxide concentrations approximately equalled the reduction in oxygen. The storage atmosphere was generally maintained in the range of 5–10 per cent carbon dioxide and 16–11 per cent oxygen.

Further research showed that low oxygen levels were of benefit but a higher level of carbon dioxide was not required. Systems were then developed with a more gas-tight storage chamber, and an external generator that burnt fuels such as petroleum gas was attached to rapidly reduce the initial level of oxygen in the store. Excess carbon dioxide produced by the generator and from ongoing produce respiration was removed by recirculating the storage atmosphere through a scrubber. The controlled admission of some air prevented the oxygen level becoming too low during storage. These developments allowed an effective atmosphere, especially for many apple cultivars, containing 2–5 per cent carbon dioxide and 2–3 per cent oxygen, to be quickly established and maintained for long periods. However, such an atmosphere needs to be constantly monitored and remedial action taken to prevent undesirable low and high levels of oxygen and carbon dioxide, respectively.

There is now considerable interest in atmosphere control in the long distance transport of perishable produce in containers. One of the factors responsible for this interest is the availability in some countries, notably the USA, of cheap liquid nitrogen. Atmosphere control in large containers has involved either the continuous introduction of carbon dioxide or nitrogen gas during the journey, or charging the container with the appropriate atmosphere before the journey, with no further introduction of gas. Carbon dioxide and nitrogen from pressurised cylinders are used, depending on whether the requirement is for high carbon dioxide or low oxygen, or both.

The advent of pressure swing adsorption machines and hollow fibre systems for separating oxygen and nitrogen in air by differential diffusion across a membrane has provided a new means of producing low oxygen atmospheres for continuous ventilation of produce in large containers or in fixed storage rooms. An advantage of these techniques is that the starting raw material, normal air, is freely available, and no hazardous gaseous by-products are generated.

The use of liquid nitrogen as a refrigerant in the transport of perishables stimulated interest in the response of fruit and vegetables to very high nitrogen levels and consequent concentrations of oxygen of 1 per cent or less, both under refrigeration and at higher temperatures. It is now known that many fruits and vegetables can withstand such atmospheres for a short period without harm and show a long-term reduction in respiration when returned to storage in air. The beneficial effects are particularly notable for non-climacteric fruit and vegetables. This finding also meant that using liquid nitrogen refrigeration for road and rail transport vehicles was unlikely to damage most perishables, because the likelihood of maintaining oxygen-free atmospheres was remote and even 1 per cent oxygen was enough for most commodities to remain viable for several days.

While carbon dioxide is generally considered non-toxic, a safe limit for exposure in the workplace has been set at 5000 μL/L; that is, 0.5 per cent. Care needs to be taken when entering rooms or containers that have been maintained under a controlled atmosphere, or when working in confined spaces adjacent to such facilities.

Storage in plastic films

Polyethylene film bags are a relatively cheap but widely available packaging material that can be used as a horticultural box liner, with the bag being sealed or remaining unsealed. Unsealed or perforated bags are commonly used to minimise weight loss and reduce abrasion damage, while sealing the bag will also generate a modified atmosphere. A major problem with sealed bags is that the atmosphere inside depends on the temperature, because the permeability of the film to gases is virtually independent of the temperatures at which produce is normally handled, whereas respiration is temperature-dependent. Thus sealed bags are risky when the temperature varies more than a few degrees, unless the produce has a low rate of respiration, or is tolerant to atmospheres that vary widely in carbon dioxide and oxygen concentration (like the banana), or both. The film commonly used is 40 μm (0.0015 inch) low-density polyethylene. To avoid carbon dioxide injuries, sachets of fresh hydrated lime (calcium hydroxide) can be included in the bag (100–200 g per 10 kg of produce).

The attainment of a modified atmosphere in polyethylene bags filled with produce can be accelerated by evacuating the bags to between 50 and 85 kPa (380–635 mm mercury) and then sealing. Since the polyethylene film is permeable to nitrogen, oxygen and carbon dioxide, the pressure inside returns to atmospheric pressure, but the initial rapid reduction of oxygen concentration is often useful. Eventually the composition of the atmosphere approaches that in bags not subjected to initial evacuation. Fruits must be removed from the bags to achieve normal ripening. If held for long periods under modified atmosphere, fruits may not ripen satisfactorily after removal.

The interest in plastic films to generate modified atmospheres has accelerated in recent years through the availability of films with greater flexibility in gas permeability. The newer films remove many of the risks of modified atmosphere storage and are often marketed as 'smart' or 'active' packaging. The types of film will be discussed in Chapter 12.

Hypobaric storage

Hypobaric storage is a form of controlled atmosphere storage. Produce is stored in a partial vacuum and the chamber is vented continuously with water-saturated air to maintain oxygen levels and minimise water loss. Ripening of fruit is retarded by hypobaric storage, due to the reduced partial pressure of oxygen and for some fruits also to the reduction in ethylene levels. A reduction in air pressure to 10 kPa (0.1 atmosphere) is equivalent to reducing the oxygen concentration to about 2 per cent at normal atmospheric pressure. Hypobaric stores are expensive to construct because of the low internal pressures required, which tends to limit the application of hypobaric storage to high-value produce.

Ethylene

Effects on fruit and vegetables

The commencement of natural ripening in climacteric fruits is accompanied by an increase in ethylene production (Chapter 3). Treatment of pre-climacteric fruits with exogenous ethylene advances the onset of ripening. This response is used widely in commercial practice to achieve controlled ripening of fruits such as banana, which is picked and transported in a mature but unripe state and ripened just before marketing (Chapter 11). The action of ethylene must, however, be avoided for such fruit during storage and transport to prevent premature ripening. In contrast, the effect of ethylene on non-climacteric fruit and vegetables offers no general commercial benefit but will reduce postharvest quality by promoting senescence, as evidenced by loss of green colour, change in texture and flavour, and by enhancement of low temperature injuries and microbial decay. However, in some instances, such as in controlled degreening of citrus fruit, these effects can be beneficially utilised (see Chapter 11).

While the levels of ethylene that trigger ripening have been well researched for most climacteric fruits, the threshold concentration that enhances senescence in non-climacteric fruit and vegetables is less well documented. A concentration of 0.1 μL/L is often cited as the threshold level, but studies in Australia indicate that the threshold level of ethylene is less than 0.005 μL/L. For practical purposes, this means there is no safe level of ethylene, and hence any reduction in ethylene concentration will bring some extension in postharvest life.

Ethylene in a storage or transport container may come from produce or from outside sources. Often during marketing, several commodity types are stored together, and under these conditions ethylene given off by one commodity can adversely affect another. Coal gas, petroleum gas and exhaust gases from internal combustion engines contain ethylene, and contamination of stored produce by these gases may introduce sufficient ethylene to initiate ripening in fruit and promote deterioration in non-climacteric produce and ornamentals. The level of deleterious action will depend on the concentration of ethylene that accumulates and the duration of exposure.

In addition to delaying ripening or senescence through reducing ethylene concentrations around produce, the sensitivity of produce to ethylene may be lessened by storing it at low temperature, and by either raising the level of carbon dioxide or decreasing the level of oxygen. Under these conditions the amount of ethylene required to induce ripening is increased. A similar effect has been demonstrated for some non-climacteric produce; for example, the ethylene-induced breakdown of chlorophyll in broccoli is less sensitive to ethylene at low temperatures.

Effects on ornamentals

Many ornamental crops are sensitive to ethylene. Their response to ethylene can be classed into growth, abscission and senescence responses. An example of an ethylene-induced growth response is epinastic curvature of poinsettia leaves and bracts. Abscission is a far more widespread response across the broad range of ornamental species. Many types of organs may abscise, including stem segments, leaves, fruit, whole inflorescences, buds and flowers, and petals. For example, fruit, leaves and stem segments of Christmas mistletoe sprigs (*Phoradendron tomentosum*) all separate upon exposure to ethylene. Ethylene-induced accelerated senescence is characterised by premature discolouration and wilting of flowers, such as carnations and cymbidium orchids.

Individual species vary widely in their relative sensitivity to ethylene, and, in general, cut flowers and flowering pot plants tend to be more ethylene sensitive than foliage lines. Among cut flowers, carnation and delphinium are considered to be very sensitive to ethylene, while gerbera and tulip are considered relatively insensitive. Among flowering pot plants, hibiscus is classed as highly sensitive and chrysanthemum is of low sensitivity. Finally, among foliage plants, schefflera is highly sensitive and nephrolepis is insensitive. However, it must be noted that ethylene sensitivity can vary markedly among genotypes (e.g. species) within a genus.

Similarly, ethylene production can vary widely between genotypes. For example, some carnations produce a marked ethylene climacteric peak during senescence, whereas others do not produce significant amounts of ethylene. As has been observed for genotype, factors that influence phenotype, including preharvest temperature, RH, light and nutrition regimes, can also affect the relative sensitivity of ornamentals to ethylene.

Methods for reducing ethylene concentrations

Avoidance of ethylene accumulation

Reduction of ethylene levels in storage rooms can be achieved by good housekeeping; that is, by storing ripe and unripe produce in separate rooms, regularly removing all rotted or damaged produce, and ensuring that natural gas pipes and cylinders, and exhaust gases from internal combustion engines, are kept well away from storage rooms.

A simple physical method to minimise ethylene accumulation is to ensure good ventilation of the storage chamber with air from outside the storage complex. The ethylene concentration in the atmosphere is normally less than 0.005 μL/L unless there is contamination from nearby industrial sources or heavy automobile traffic. Ventilation with external air could be applicable where no large temperature differential exists between the external air and the storage chamber, provided it was at an appropriate relative humidity. Increased water loss from produce can be an issue if the humidity of the external air is low. If there is a large temperature difference it may be necessary to cool the air before admitting it to the chamber. The use of recently developed compact but efficient heat exchangers has resurrected interest in ventilation for use in cool stores, and commercial ventilation units are now on the market.

Oxidation with potassium permanganate

Ethylene in the atmosphere can be oxidised to carbon dioxide and water using a range of chemical agents. Potassium permanganate is quite effective in reducing ethylene levels. Since it is non-volatile, potassium permanganate can be physically separate from produce, thus eliminating the risk of chemical injury. To ensure efficient destruction of ethylene, a large surface area of potassium permanganate is achieved by coating an inert inorganic porous support, such as alumina or expanded mica, with a saturated solution of potassium permanganate. Potassium permanganate used in this manner has been found to retard the ripening of many fruits. Table 6.2 demonstrates the benefit to be obtained with banana when potassium permanganate is used in conjunction with modified atmosphere storage in polyethylene bags. The high carbon dioxide and low oxygen atmosphere generated within the sealed bags decreases the response by bananas to ethylene while the addition of potassium permanganate further retards ripening by maintaining ethylene at a low level for a long period.

The presence of high humidity in storage containers limits the longevity of potassium permanganate since it also reacts with water. While the extent of commercial use of potassium permanganate alone or in conjunction with modified atmosphere is difficult to document, the Australian navy uses tubes of potassium permanganate in fruit and vegetable chambers on board its vessels, and a range of retail products designed for use in domestic refrigerators are being marketed.

Table 6.2 Shelf life of banana held at 20°C

Treatment	Shelf life (days)
Air	up to 7
Sealed polyethylene bags	14
Sealed bags + potassium permanganate	21

SOURCE Derived from K.J. Scott, W.B. McGlasson and E.A. Roberts (1970) Potassium permanganate as an ethylene absorbent in polyethylene bags to delay ripening of bananas during storage. *Australian Journal of Experimental Agriculture and Animal Husbandry* 10, pp. 237–40.

Oxidation with ozone

Ozone (O_3) is a suitable oxidising agent for destroying ethylene and is generated readily from atmospheric oxygen by an electric discharge or ultraviolet radiation; since it is gaseous, it readily mixes with ethylene. The actual oxidant species is probably a combination of ozone and atomic oxygen, a highly reactive free radical formed from ozone. Some precautions must be taken with ozone: it is a reactive substance and will corrode metal pipes and fittings in refrigeration equipment and react with paper products used to package produce. In addition, ozone readily injures produce and can be toxic to humans at relatively low concentrations – 0.1 μL/L has been determined as the limit of exposure in the workplace. The widespread use of ozone has been hampered by difficulties in controlling its concentration. These problems with ozone can be overcome by using it in a recycling system as depicted in Figure 6.2; ozone is generated in a container with ultraviolet radiation, air contaminated with ethylene is passed through the chamber, where the ethylene is oxidised, and excess ozone is removed by reduction on a substrate such as steel wool. Some small scale commercial ultra-violet scrubbers have been produced but are not in widespread use.

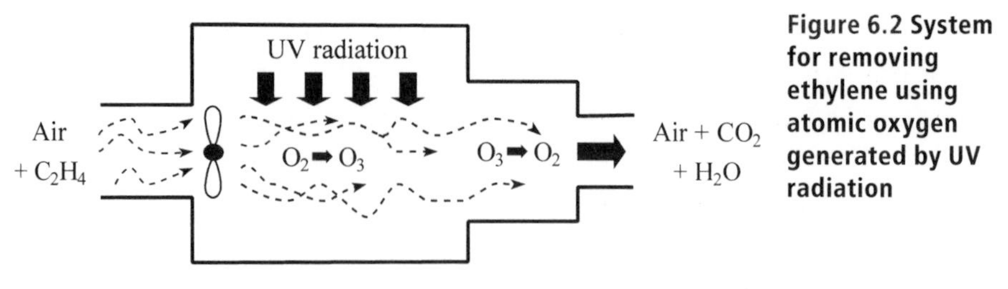

Figure 6.2 System for removing ethylene using atomic oxygen generated by UV radiation

SOURCE Derived from K.J. Scott and R.B.H. Wills (1973) Atmospheric pollutants destroyed in an ultraviolet scrubber, *Laboratory Practice* 22, pp. 103–6.

Other oxidants

Activated charcoal that has been brominated will effectively oxidise ethylene but its potential health hazard is that it generates bromine gas when in contact with excess water. Recent laboratory studies have identified a range of chemicals, such as the tetrazines, which react specifically with ethylene. Their greater specificity for ethylene makes them relatively more efficient in ethylene scavenging than general oxidising agents such as potassium permanganate. Commercial applications are focusing on including them in plastic packaging films. A problem to be overcome is the instability of tetrazines in the presence of moisture.

Other chemicals

A range of chemicals that are potentially toxic to humans can be used as anti-ethylene treatments with ornamentals, because they are not eaten. Most notably, cut stems of ornamental material can be pulsed with silver ions as silver thiosulphate (STS). Whole plants or plant parts can be dipped or sprayed with STS. Silver ions block the ethylene binding site, thereby preventing its action. Silver is a toxic heavy metal and there is considerable environmental concern arising from the disposal of waste dip solutions and treated produce. This concern has prompted the search for safer alternative treatments.

1-Methylcyclopropene (1-MCP) is a potential replacement for silver. It also acts by blocking ethylene binding sites. 1-MCP is presently applied as a gas, but work is underway to develop a liquid formulation of this chemical. 1-MCP is considered safe in terms of having low or no mammalian toxicity and low or no adverse environmental impact. It has also been widely evaluated as a postharvest treatment for a broad range of fruit and vegetables. It is registered for commercial use as SmartFresh™ on both temperate and tropical crops in many countries. Current commercialisation of 1-MCP has tended to focus on extending the storage life of apple fruit, particularly through a delay in softening. In the case of other climacteric fruit, treatment with 1-MCP can provide significant extension in display life but there are reports of poorer flavour and aroma in some fruits. This may be due to suppression of the climacteric.

Treatment with ethylene synthesis inhibitors is an alternative, or perhaps an adjunct, to treatment with ethylene binding site blockers. Two inhibiting compounds are aminoethoxyvinylglycine (AVG) and aminooxyacetic acid (AOA). Both are used after harvest to treat ornamentals, such as carnation, against ethylene. Preharvest application of AVG (Retain™) is made to apple fruit, allowing them to reach greater maturity and colour prior to harvest. From a postharvest perspective, a potential drawback of ethylene synthesis inhibitors is that they only confer protection against endogenously (internally) produced, and not exogenous (external), ethylene.

Another approach to reducing or eliminating ethylene effects on ornamentals and other crops is genetic engineering. Non-edible genetically engineered ornamentals are likely to be accepted by the general public ahead of genetically engineered fruit and vegetables. Carnation genotypes that formerly produced ethylene during senescence (i.e. climacteric characteristic) have been genetically engineered using anti-sense gene technology to produce very little ethylene. These genetically engineered carnations, which display non-climacteric senescence characteristics, have a longer vase life. Similarly, carnation genotypes with blocked ethylene binding and signal transduction and translation can be engineered. These genetic modifications emulate the effects of the chemical ethylene synthesis (e.g. AVG) and binding (e.g. 1-MCP) inhibitors, respectively.

Other gases

Carbon monoxide (CO) is not generated by fresh produce, but it may be introduced to storage atmospheres by equipment powered by internal combustion engines. Carbon monoxide may reach levels toxic to persons working in the storage chambers, and some produce may respond in a way that mimics the effects of ethylene. However, there are examples of beneficial responses to added carbon monoxide; for example, the control of butt discolouration and retarded growth of *Botrytis* rots in lettuce. Adding about 5 per cent carbon monoxide to containers or pallet loads of various perishable fruits held in controlled atmospheres is now considered advantageous and is used to a limited extent for export shipments and long-distance land transport of specific produce.

Two oxides of nitrogen, nitrous oxide (N_2O) and nitric oxide (NO), have been reported to extend the postharvest life of a range of horticultural produce, albeit by vastly different methods. Nitrous oxide is a stable gas that is used at quite high concentrations (20–80%) as a replacement for nitrogen in the storage atmosphere. Nitrous oxide is presumed to suppress metabolism in a similar way to carbon dioxide, as both compounds have a similar structure, but nitrous oxide is tolerated by produce at much higher concentrations. The postharvest benefit of nitrous oxide is, however, only achieved while the nitrous oxide is present. Nitric oxide, on the other hand, is a highly reactive free radical gas and is applied as a short-term treatment (hours) at relatively low concentrations (parts per million), which has a long-term benefit for produce. Its mode of action is not yet known but is expected to be multi-faceted, in parallel with the many effects nitric oxide has been found to have on mammalian physiology. Nitric oxide has a potential advantage in that it is a natural metabolite of both mammals and plants.

Some other organic volatiles, such as acetaldehyde and ethanol, which are generated by produce, may have a role in quality maintenance. This is currently under investigation, particularly in relation to antimicrobial properties for certain compounds. Ethanol vapour has also been shown to markedly reduce the incidence of superficial scald, a physiological disorder in apples, so it could be an alternative to current dipping treatments, which raise some health and environmental issues.

7 Technology of storage

Three principles govern the preservation of fresh fruit, vegetables and ornamental plants: they keep better when cold, they are damaged by freezing, and they shrivel or wilt in dry air. Equipment and systems for controlling the temperature and RH of the air around produce are applications of these principles. An additional factor that can influence the postharvest life of fresh produce is the composition of the storage atmosphere, particularly the concentrations of oxygen, carbon dioxide and ethylene. This has led to controlled or modified atmosphere storage (Chapter 6).

Methods of storage

In-ground storage

Pit storage, or clamp storage, is a simple, low technology on-farm technique that is still beneficially practised in some countries. Hard vegetables, such as potato, turnip and late season cabbage, are piled into pits dug into a hillside or in some other well-drained position. The pits are lined with hay or straw; the produce is also covered with straw, followed by 10–20 cm of sods and earth to protect it against freezing or from excess heat and to deflect rain. The inclusion of piped ventilation to the outside, to reduce respiratory self-heating, is an advantage. A clamp has been shown to be a suitable method of storing cassava for up to two months in the tropics (Figure 7.1). In Europe, perishable produce was traditionally stored in cellars and caves that were cooler than above-ground buildings in warm weather,

Figure 7.1 Cassava storage clamp

(Courtesy of R.H. Booth, Food and Agriculture Organization of the United Nations, Rome.)

and warmer in winter. These methods are still practised in some parts of the world. Effective drainage and protection from rain are essential. Cellar performance is improved by providing controlled ventilation openings to allow the entrance of cold air and the exit of warm air by convectional circulation when cooling is required. Although temperatures are generally not optimal, a good cellar will provide satisfactory storage for hard vegetables and long-keeping fruits such as apple.

Air-cooled stores

These are simple, insulated structures above ground, or partly underground, which are cooled by circulation of colder, outside air. When the temperature of the produce is above the desired level, and if the temperature of the outside air is lower (generally at night), air is circulated throughout the store by convectional or mechanical means through bottom inlet vents and top outlets fitted with dampers. Fans may be installed and are controlled manually, or automatically with differential thermostats. The air may be humidified, a process that can also be automated. Air-cooled stores are cheap to construct and operate and are still widely used to store potato and sweet potato, both of which need relatively high storage temperatures to avoid accumulation of sugar and chilling injury, respectively.

Potatoes are commonly stored in bulk piles in stores with air delivery ducts under the floor, or at floor level, and with suitably spaced air outlets.

Bulk piles or bins of onions are also ventilated with air, which is an economical way to ensure that the outer scale leaves remain dry and free of decay. Garlic corms are similarly ventilated with air to prevent mould growth.

Ice refrigeration

An advance on air-cooled storage was the use of natural ice as a refrigerant. The lower temperatures obtained enabled longer storage of meat and other perishable foods, including fresh fruit and vegetables. In North America, and North and Central Europe, ice was harvested in the winter from frozen lakes and ponds and stored in insulated 'ice houses'. The melting of 1 kg of ice absorbs 325 kJ of heat, but the considerable bulk of ice needed and disposal of the melt water are disadvantages. The introduction of the small 'ice box' or 'ice chest' was a great advance in the domestic and small-scale commercial preservation of perishable foodstuffs. Ice produced by mechanical refrigeration has several modern commercial applications as an adjunct to refrigeration (Chapter 4).

Mechanical refrigeration

The father of modern refrigeration systems was undoubtedly the Australian James Harrison. By 1851 he had designed and built the first ice-making plant in the world, incorporating a small refrigeration compressor with its auxiliary equipment and ice tank, at Geelong in Victoria. In 1854 Harrison was granted British Patent No. 717 for 'the production of cold by the evaporation of volatile liquids in vacuo', an invention probably equal in importance to that of the steam engine. The general principles of Harrison's design remain virtually unchanged in modern refrigeration plants. The system developed rapidly, and mechanically refrigerated cold stores, insulated with natural materials, such as sawdust or cork, were operating within a few years. A shipment of frozen beef from Australia to England in 1879 was the first successful long-distance shipment of perishable food by sea; soon after, the first mechanically refrigerated cool stores for apple and pear were in operation.

A refrigeration plant consists of four basic components: the compressor, in which the refrigerant gas (carbon dioxide, ammonia or halogenated hydrocarbons) is compressed (and unavoidably heated); the condenser, either air-cooled or water-cooled, in which the hot gas is cooled and condensed to a liquid; the expansion valve; and the evaporator coils, in which the

Figure 7.2 Basic component parts of mechanical refrigeration plant. The part of the cycle from the compressor to the thermal expansion valve operates under high pressure to enable condensation of the hot gaseous refrigerant. The evaporator coils operate at low pressure to enable the refrigerant to boil. The condenser may be either air or water cooled.

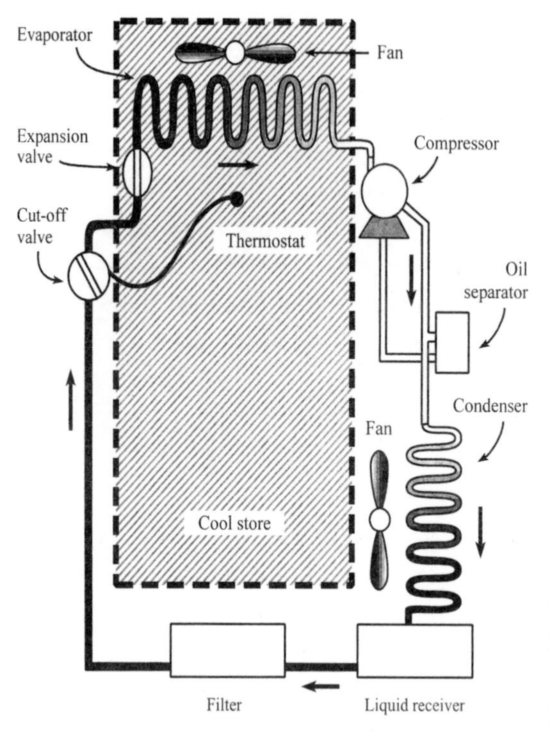

liquid is permitted to boil and so remove heat from its surroundings (Figure 7.2). Fans are usually necessary to circulate the storage air over the cooling coils of the evaporator and through the stacks of produce in the store. The main agent for transfer of heat from the interior of the store to the cooling coils is air movement, although radiation and convection may play a small part.

Removing heat from the circulating air in a cool room using evaporator coils is referred to as the 'direct expansion system'. The 'indirect expansion system' is a variation in which a liquid, usually a mixture of polyethylene glycol and water with a low freezing point, is cooled and the cold solution is circulated through the cooling coils in the cool room. The reservoir of cold polyethylene glycol solution can buffer periods of high cooling demand during the day when warm produce is being loaded. Inevitably, the surface temperature of the cooling coils must be lower than that of produce to ensure that heat from all sources in a cool room is removed and produce remains at a constant temperature. This temperature gradient is accompanied by a gradient in vapour pressure (vapour pressure deficit) between the produce and the evaporator coils. This VPD enhances water loss from produce. For produce with high transpiration rates (leafy and root vegetables, mushrooms and cut flowers) a preferred system is to cool and humidify the room air by passing it through a shower of cold water that has been cooled by mechanical refrigeration. This indirect expansion system provides air at 1–2°C and RH

Plate 1 Retail display of fruit and vegetables designed to be visually attractive to consumers

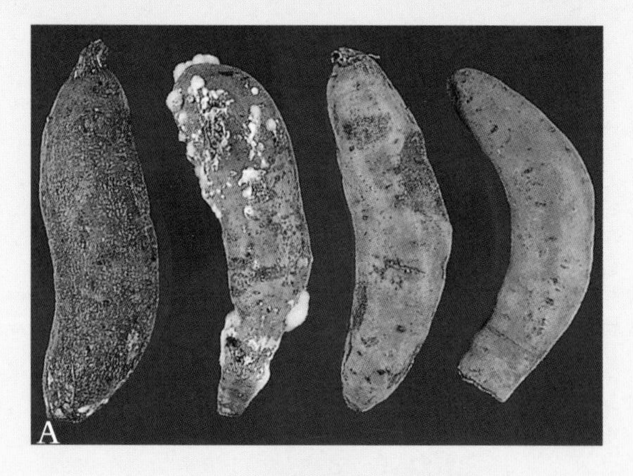

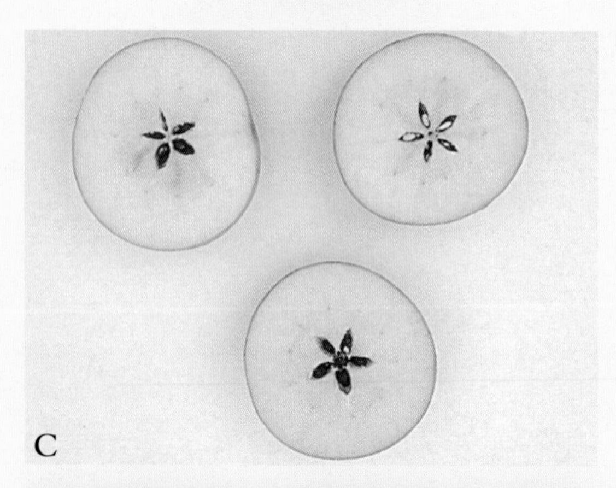

Plate 2 Illustrations of non-pathogenic diseases: (A) occurrence of chilling injury of sweet potato following 7 weeks storage at (left to right) 0°C, 5°C, 10°C and 15°C; (B) bitter pit of apple; (C) core flush of apple stored at 0°C; (D) storage spot of Valencia orange stored at 5°C; (E) blackheart of pineapple (in Australia this condition is usually caused by preharvest chilling); (F) superficial scald of Granny Smith apple stored at 0°C showing that browning does not extend below the skin; (G) chilling injury in anthurium stored at 0°C (bottom row) and 5°C (middle row) for 3 days, followed by 2 days at 20°C (the flowers in the top row were held for 3 days at 15°C followed by 2 days at 20°C); (H) chilling injury of frangipani (bottom) following 3 days at 4°C and 3 days at 20°C (the non-chilled flower [top] was held for 6 days at 22°C).

(Courtesy of B.B. Beattie, formerly NSW Department of Agriculture [plates A–E]; Dr J.B. Golding, NSW Department of Primary Industries [plate F]; and Dr D.H. Simons, formerly University of Queensland [plates G–H]).

E

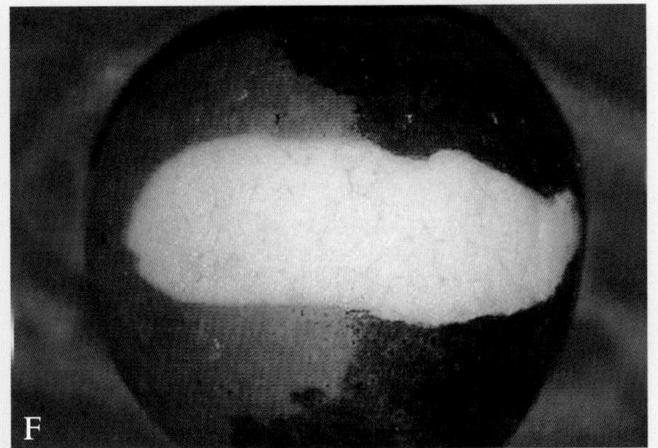
F

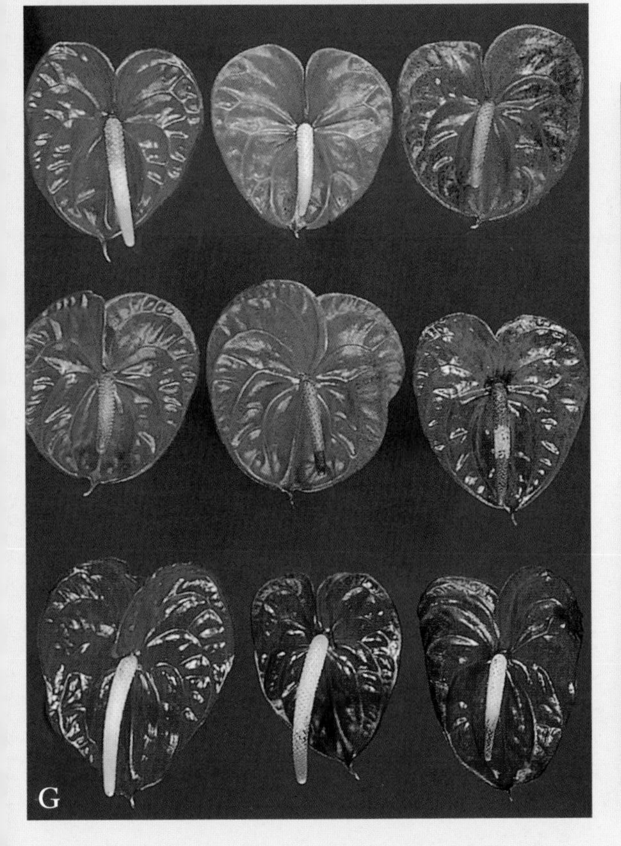
G

H

H

Plate 3 Illustrations of pathological diseases: (A) brown rot (*Monilinia fructicola*) of peach; (B) blue mould (*Penicillium italicum*) (left) and green mould (*P. digitatum*) (right) of oranges; (C) anthracnose (*Colletotrichum gloeosporioides*) of avocado; (D) crown rot of bananas caused by several species of fungi; (E) stem-end rot of mango (*Dothiorella dominicana* is the most common cause of this rot in Australia); (F) secondary fungal decay of cucumber at 20°C following chilling at 0°C; (G) Geraldton wax flowers infected with *Alternaria alternata* (top) and *Botrytis cinerea* (middle), and an uninfected flower (bottom); (H) freesia infected with *Botrytis cinerea*.

(Courtesy of B.B. Beattie, formerly NSW Department of Agriculture [plates A–E]; Dr D.H. Simons, formerly University of Queensland [plate F]); Dr Melissa Taylor, University of Queensland [plate G]; and Dr Tassos Darras, Technological Educational Institute of Kalamata, Greece [plate H].

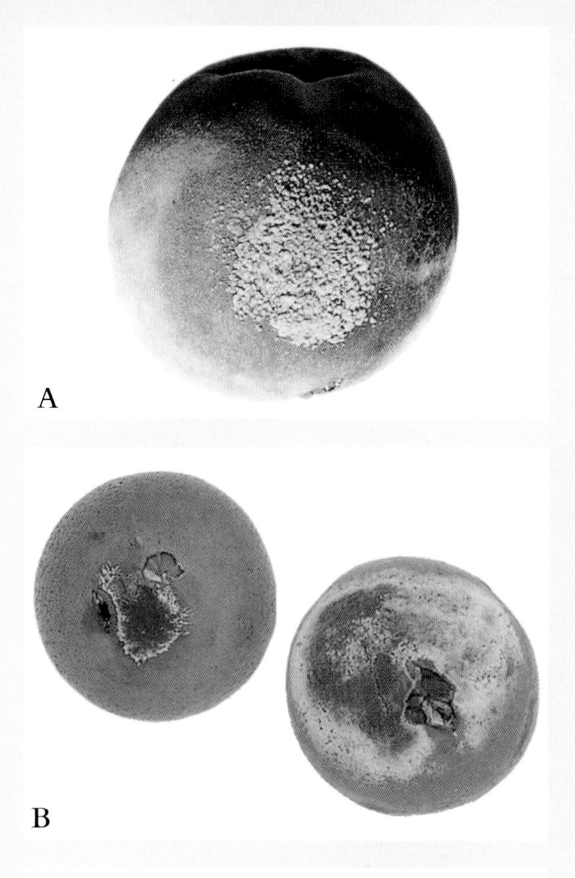

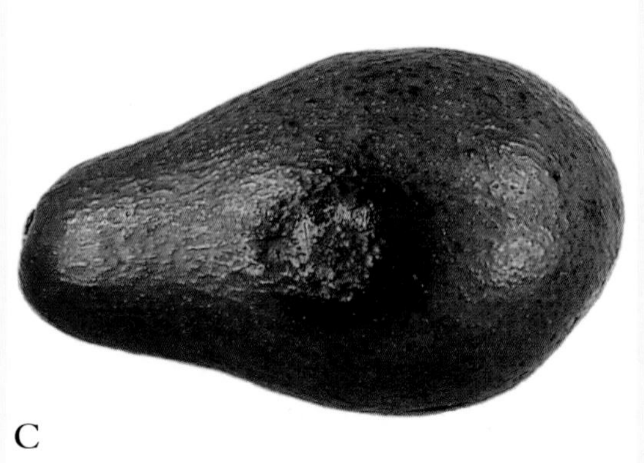

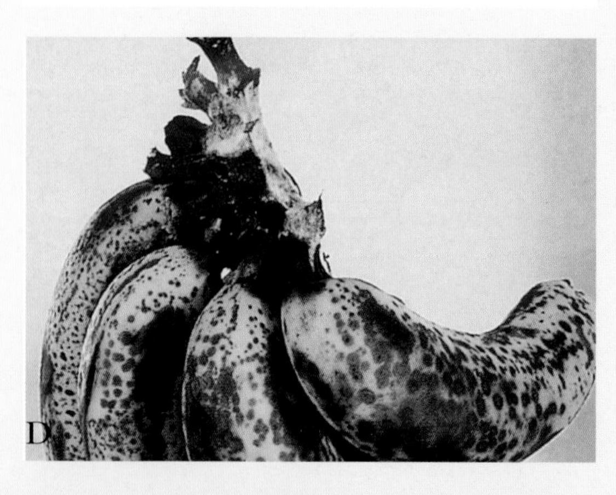

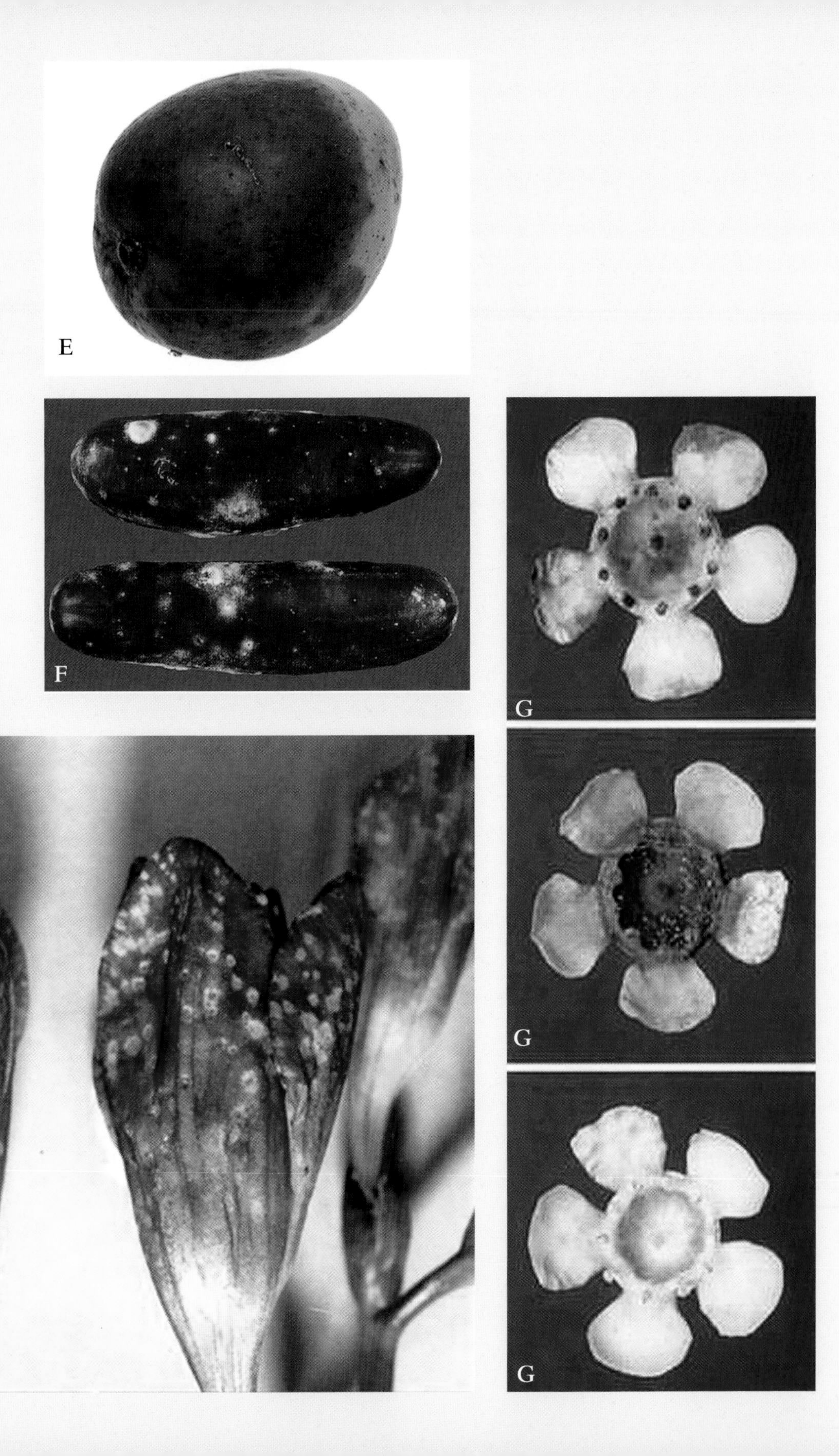
E
F
G
G
G

Plate 4 Modern machinery for sizing fruit by weight. Imaging systems are also available for grading by size, colour and skin blemishes. Near infra-red sensors can be added to grading lines to sort fruit for soluble solids concentration.

(Reproduced with permission of The Taste Factory, Cobram Victoria, Australia.)

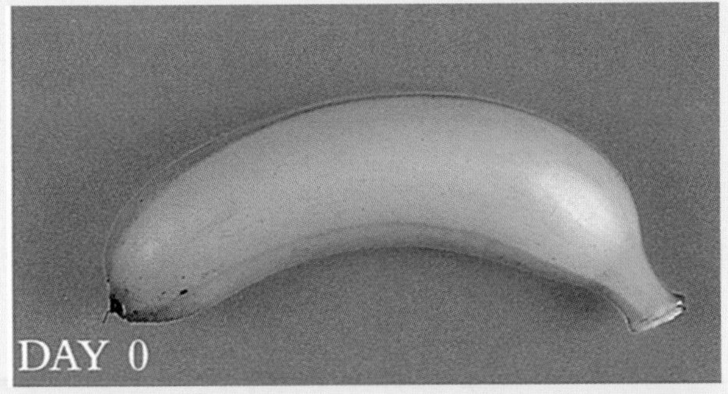

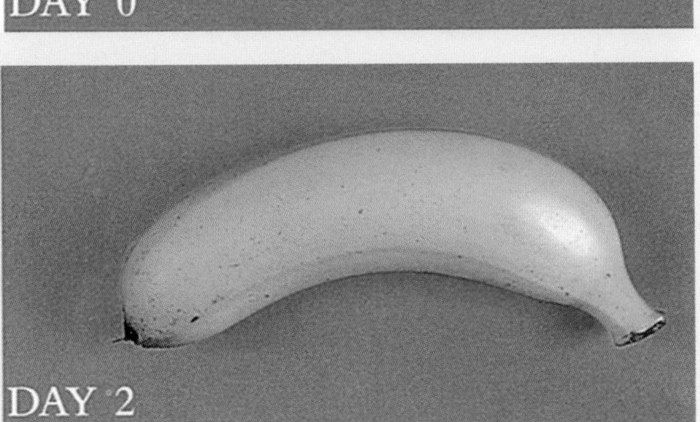

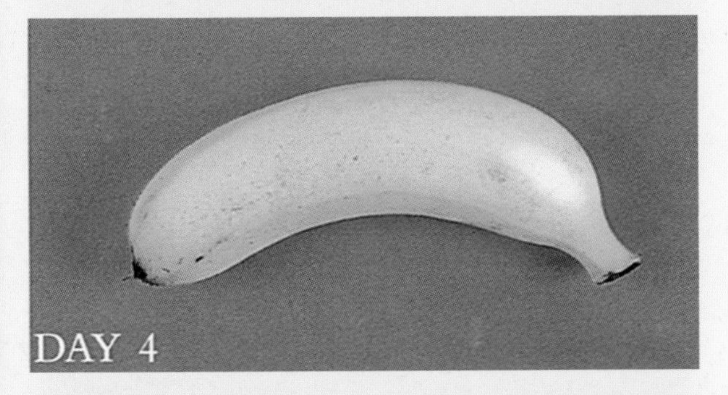

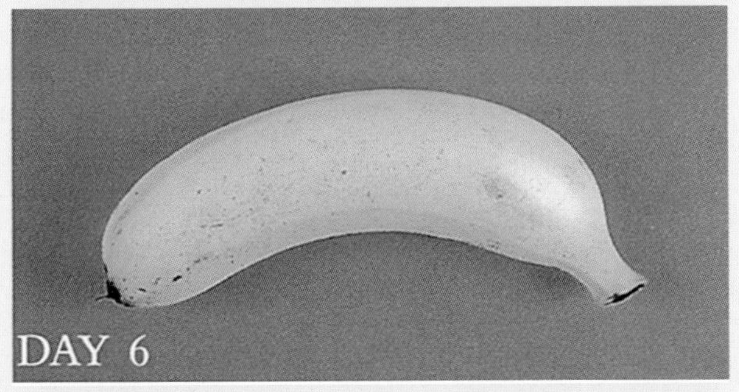

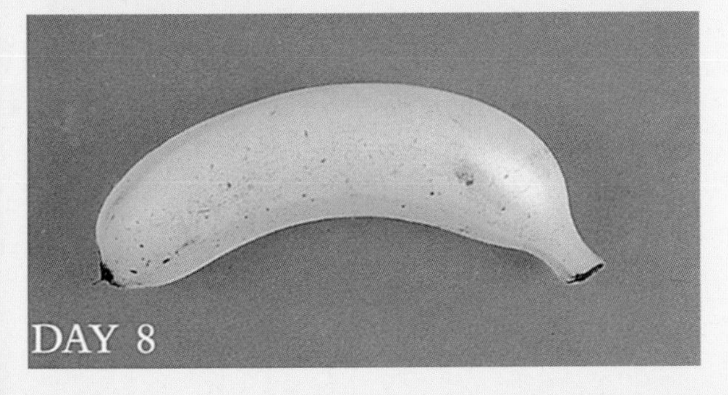

Plate 5 Ripening scale of Cavendish banana (*Musa acuminata* var Williams). The plates show the changes in colour of a single banana at 20°C at 2-day intervals; Day 0, initial colour before application of ethylene at 100μL/L for 24 hours; Days 2–8, the progressive yellowing of the fruit. Soluble solids concentration reached a maximum in comparable fruit at about Day 8.

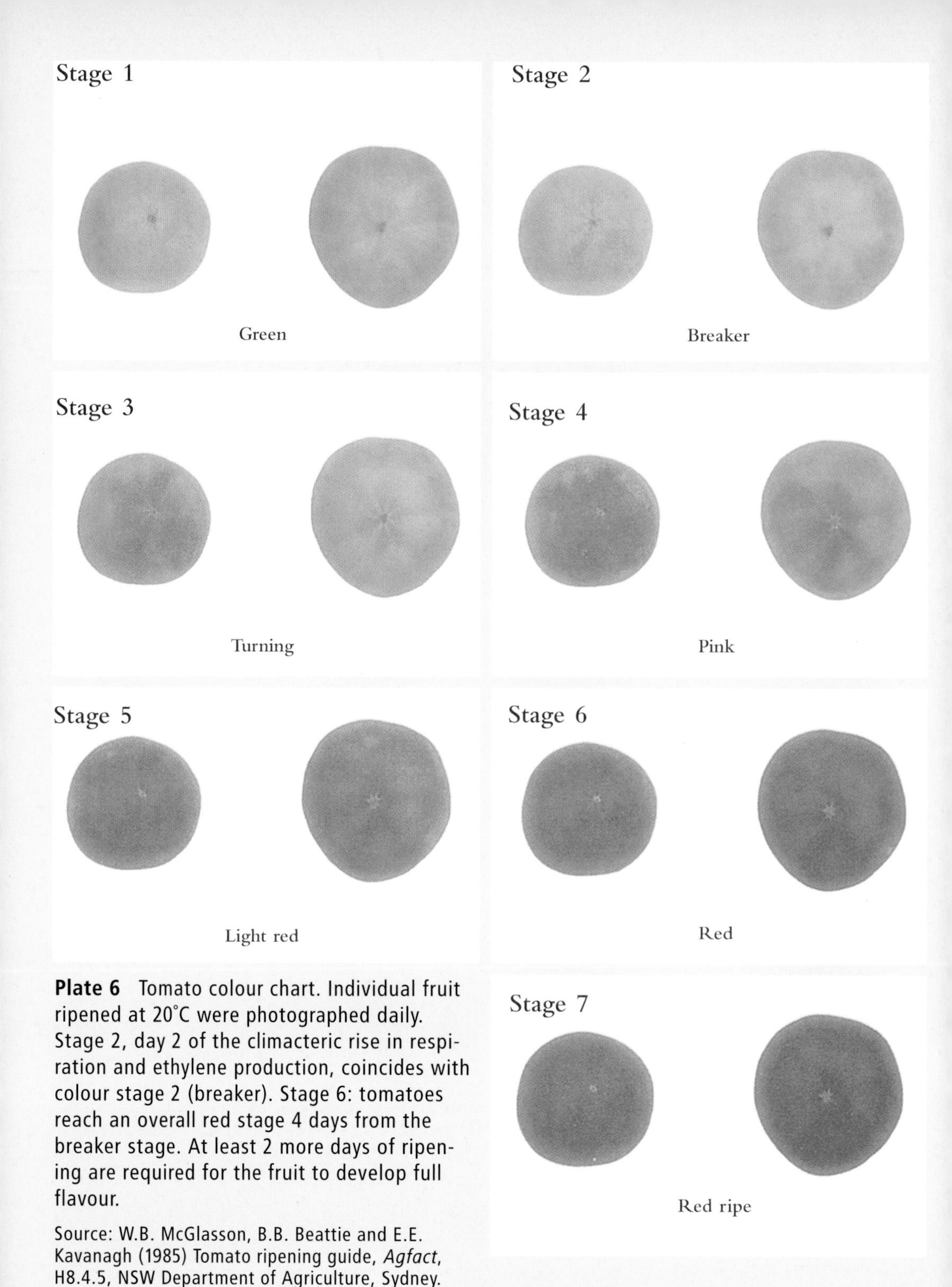

Plate 6 Tomato colour chart. Individual fruit ripened at 20°C were photographed daily. Stage 2, day 2 of the climacteric rise in respiration and ethylene production, coincides with colour stage 2 (breaker). Stage 6: tomatoes reach an overall red stage 4 days from the breaker stage. At least 2 more days of ripening are required for the fruit to develop full flavour.

Source: W.B. McGlasson, B.B. Beattie and E.E. Kavanagh (1985) Tomato ripening guide, *Agfact*, H8.4.5, NSW Department of Agriculture, Sydney.

Plate 7 Equipment for measuring soluble solids concentrations (SSC) in fruit. Representative samples of fruit tissue are crushed using a garlic press, and the extracted juice is placed on the prism of a hand-held refractometer or in the measuring cell of an electronic refractometer. The readings are referenced against sucrose solutions at 20°C and expressed as soluble solids % or °Brix.

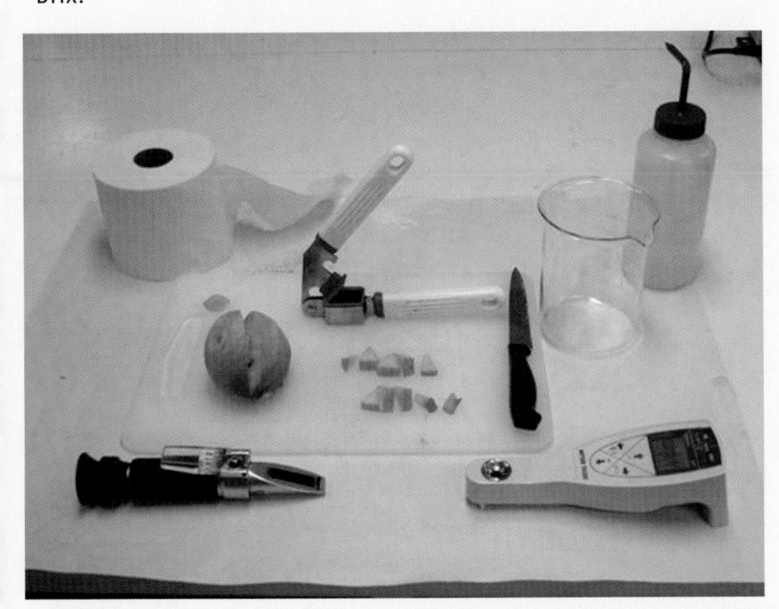

Plate 8 Non-destructive, colour enhanced, proton magnetic resonance, transverse section images of ripened Kensington Pride mango fruit. The darker areas have lower water activity and the lighter areas have higher water activity. LEFT: Fruit damaged by a heat disinfestation treatment applied when the fruit was mature-green. The dark area in the centre of the fruit is the seed, and the dark areas just inside the periphery of the fruit show heat-injured parts of the mesocarp, where starch-to-sugar hydrolysis during ripening was impaired. RIGHT: Fruit infested by a Queensland fruit fly larva, which developed from an egg oviposited when the fruit was mature-green. The dark area in the centre of the fruit is the seed, and the dark areas bounded by red areas, between the periphery of the fruit and the seed, show parts of the mesocarp where air-filled galleries arose as a result of the larva feeding on the tissue.

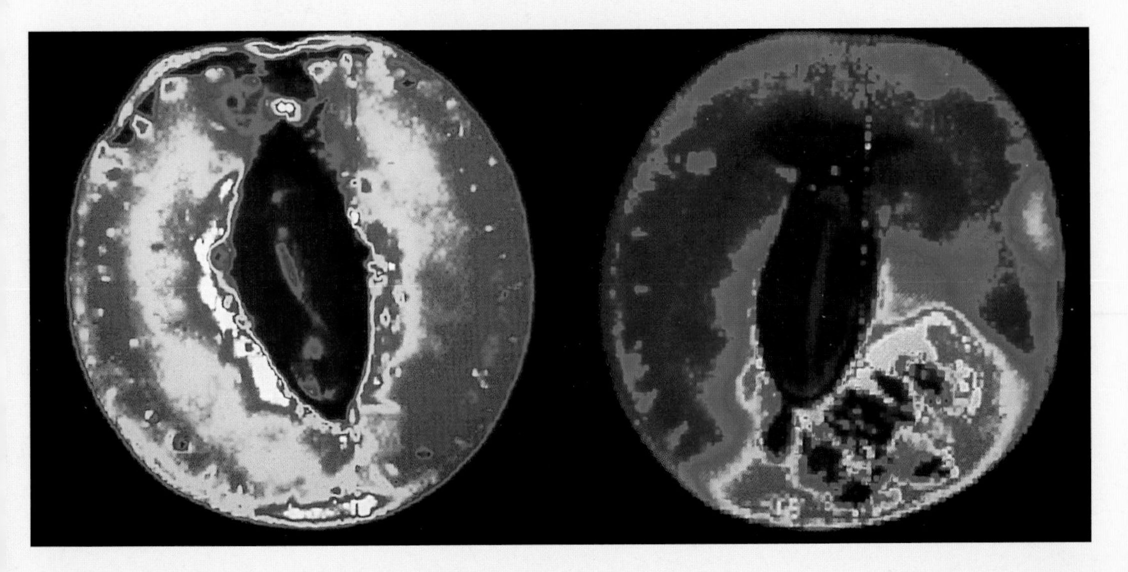

Plate 9 Sanitising fruit by spraying with an iodine solution

Plate 11 Infesting nectarines with Queensland fruit fly
(Supplied by NSW Department of Primary Industries.)

Plate 10 Some important quarantine insect pests: (A) Queensland fruit fly, *Dacus tryoni* (B) adult light brown apple moth, and (C) its pupating larva, showing the typical protective webbing (D) thrips on a cymbidium orchid flower.

(Plates A–C supplied by NSW Department of Primary Industries; plate D reproduced with permission of Australasian Biological Control Inc. from *The Good Bug Book*, 2nd ed. 2002.)

A

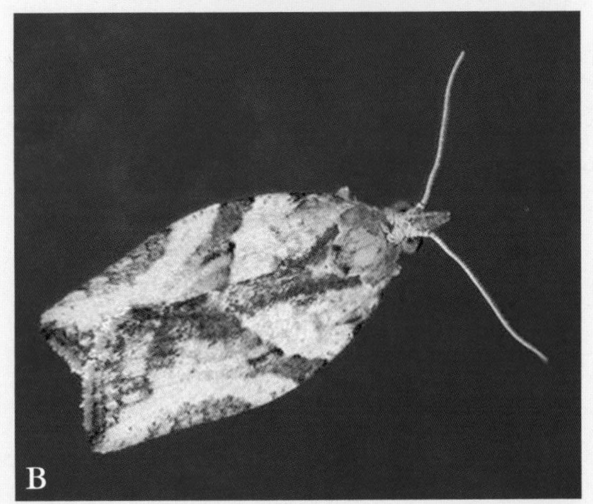

B

D

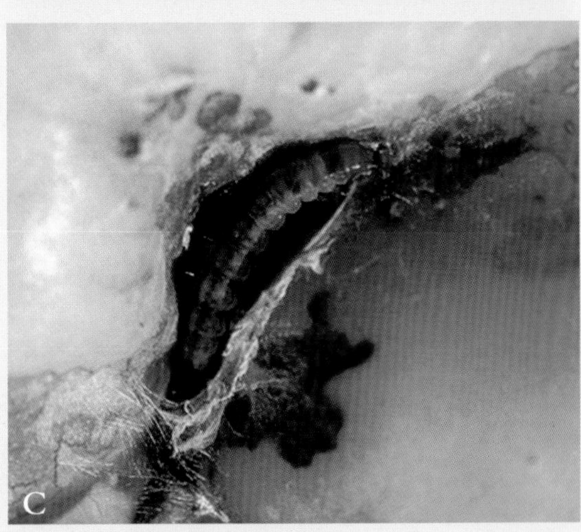

C

Plate 12 A retail display of minimally processed vegetables in a refrigerated cabinet maintained at <5°C. The sealed plastic film bags maintain a modified atmosphere around the product and are printed with information, including the recommended use-by date.

(Used with permission of Woolworths Ltd.)

Plate 13 Onions and potatoes in plastic mesh bags

Plate 14 Display of fruit and vegetables in a self-service retail outlet

(Used with permission of Woolworths Ltd.)

of more than 98%. A variation of this system allows accumulation of ice on the evaporator coils (ice bank) to augment cooling capacity at times of high demand and to take advantage of lower off-peak electricity charges if this concession is available.

Design and construction of cool stores

A cool store is a large, thermally insulated box, with doors for entry and some means of cooling the interior. Cool stores for horticultural produce have special requirements in comparison with other refrigerated stores. These include a high cooling capacity, close control of temperature, and RH of greater than 90%. A common minimum design criterion is to provide capacity to cool a daily intake of 10 per cent of the capacity of the store at an initial rate of not less than 0.5°C per hour. Such a capacity requires 1 tonne (3.5 kW) of refrigeration capacity per 18 tonnes of produce for small stores of up to about 150 tonnes capacity, and about 1 tonne per 25 tonnes for larger stores. The capacity of larger stores can be varied by having two or more compressors or by the technique of cylinder unloading in one compressor.

Accurate temperature control requires spatial variation of no more than ±1°C and a variation in time in any one position of no more than ±0.5°C. A temperature difference of 1°C over the storage period has significant effects on most produce, especially those stored at less than 5°C. The optimum thickness of insulation in walls and ceiling needs to keep the overall heat transfer to about 0.3 kJ m^{-2} h^{-1}. This gives about the most economical ratio of cost of refrigeration capacity to cost of insulation and also enables high RH to be maintained. The best insulation is the cheapest that will perform as required. Floors generally require half the thickness of insulation that is used in the walls. A vapour barrier of thick polyethylene, laminated foil material, or the equivalent, having a low water-vapour transmission rate is placed on the warm side of the insulation, to prevent moisture migrating to, and condensing within, the insulation.

Cool stores may be constructed in many ways, and provided that the above conditions are met, all can be satisfactory. Many modern cool rooms are either sandwich panel construction with polystyrene foam slabs as the insulation within the prefabricated panels, or foamed-in-place polyurethane is applied to the inner faces of the structure. The 'skins' on the outsides of the insulation are made of metal, commonly aluminium or zinc-coated steel, or waterproof plywood (Figure 7.3). Floors are constructed of reinforced concrete capable of carrying point loads from forklifts as well as stacking loads. Cooled air is generally supplied by forced or induced draft coolers (FDCs or IDCs) (Figure 7.4), consisting of framed and finned closely spaced cooling coils, fitted with fans to circulate the air over the coils. Some means of defrosting the coils is also required when storage temperatures are low and the coil surface operates at temperatures below 0°C.

Design and construction of controlled atmosphere stores

Large-scale, long-term CA storage was established with apples and pears but is now also practiced for kiwifruit and to a limited extent for speciality products such as sweet or low-pungency onions. Initially, storage room ventilation was restricted to allow carbon dioxide from fruit respiration to accumulate to the desired level. Atmospheres in such stores typically contained 5–10 per cent carbon dioxide and 16–11 per cent oxygen, as one part carbon dioxide is produced by the fruit for each part oxygen consumed.

Figure 7.3 A modern cool room being constructed using metal-clad panels insulated with polystyrene. A Colorbond finish has been applied to the panels to facilitate cleaning. All joints and entry points for electrical cables and plumbing are sealed with waterproof silicone mastic to prevent water entering the insulation.

Figure 7.4 A cool room in which the air is blown through the coils by fans mounted at the back of the coils (forced draft cooling). The more common arrangement is to mount the fans in front of the coils (induced draft cooling), which gives better air distribution. In this example, the palletised produce is placed on racks.

When research revealed that an atmosphere of 2–3 per cent carbon dioxide with 2–3 per cent oxygen was suitable for many apple and pear cultivars (Chapter 6), a much more gas-tight room was required. This required highly specialised methods of construction. Furthermore, ventilation of the store with outside air to control the carbon dioxide concentration was not possible as it would introduce too much oxygen; therefore, some means of absorbing, or 'scrubbing out', the excess carbon dioxide was required. Relatively simple methods were developed, by which carbon dioxide was adsorbed physically or absorbed chemically with dry hydrated lime. These CA stores required the addition of equipment to measure and control the concentrations of both carbon dioxide and oxygen. Being a sealed chamber, the refrigeration system has to be completely reliable, and the room has to be fitted with accurate and reliable, remote-reading thermometers.

Modern CA stores employ gas separators such as pressure swing adsorption (Figure 7.5) or hollow-fibre membrane systems (Figure 7.6) that can generate gas streams containing low oxygen concentrations by separating oxygen and nitrogen from air. These separators have the major advantage of producing clean atmospheres. Oxygen concentrations can be reduced rapidly by flushing the cool room, and once the required oxygen levels are reached, carbon dioxide concentrations are controlled by scrubbing with activated charcoal adsorbers that are also operated in a pressure swing mode. These modern carbon dioxide scrubbers have replaced the more cumbersome system of using hydrated lime in paper sacks. The operation of low-oxygen stores is now relatively simple and is frequently automated. Despite these advances, it is imperative to build gas-tight stores to ensure that separators are used economically. These land-based systems for generating and maintaining CA have been adapted to refrigerated ship holds and refrigerated shipping containers. Low-oxygen atmospheres can also be achieved by flushing storage rooms or shipping containers with liquid or compressed nitrogen in locations where nitrogen is relatively cheap.

Figure 7.5 Pressure swing adsorption machine. Filtered dry compressed air is passed alternately through two chambers packed with molecular sieves that retain oxygen when under high pressure and allow compressed nitrogen of at least 99% purity to be generated. When the adsorption capacity in one chamber is nearly reached, the compressed air is automatically switched to the second chamber and the first is purged at atmospheric pressure to release the adsorbed oxygen.

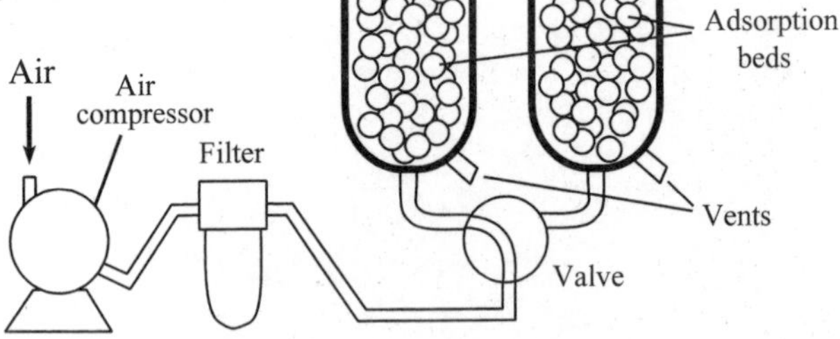

Figure 7.6 Hollow-fibre membrane system. High-pressure dry, oil-free air is passed through hollow fibres constructed from a polymer that is differentially permeable to gases. Oxygen, carbon dioxide and ethylene permeate the fibres rapidly and are purged to the atmosphere, enabling the generation of a high-pressure stream containing up to 99.5% nitrogen.

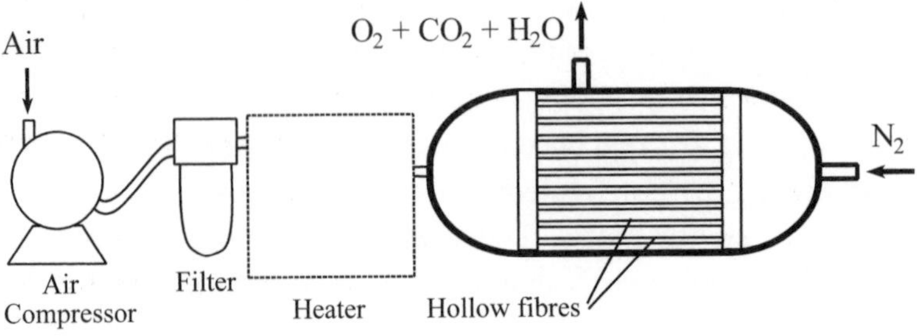

Construction

An essential feature of a CA store is the provision of an effective gas barrier, which is most conveniently placed directly on the inside of the insulated surface. If the external vapour barrier is defective, however, moisture that penetrates this barrier will then be contained on the inside of the gas barrier, leading to water-logging and destruction of the insulation. Application of foamed-in-place polyurethane enables the satisfactory conversion of existing cool stores to CA and also provides a cheap method of construction by completely lining the interior of a simple metal chamber. If correctly applied, the polyurethane provides both insulation and a gas barrier. A pressure relief device (burp valve), usually a water trap, fitted through the walls of such a rigid, gas-tight structure is essential to avoid damage by limiting pressure differentials to 370 Pa (15 mm water gauge).

Controlled atmosphere stores are lethal

While CA stores support plant life at a low level, they will not support mammalian life. CA stores should be treated with caution to ensure that no one is ever exposed to such an atmosphere, unless wearing an efficient respirator with its own oxygen supply. Chambers in transport vehicles where the atmosphere has been substantially modified are especially dangerous and, like CA stores, must be well ventilated before entry.

Produce storage management

Produce will come out of storage at a high quality only if it was of high quality on entering the store, and if the storage facilities are managed to a high standard. Given correct selection and handling of produce, the success of subsequent storage depends on quickly reducing the temperature of produce to the desired level and maintaining it with little variation; closely monitoring humidity and gas concentrations in the storage atmosphere; and avoiding over-storage.

Precooling the rooms

Storage rooms are generally brought down to the appropriate temperature a few days before produce intake commences. Three days is enough for a fully insulated room, but rooms without floor insulation should be precooled for a week to ensure that the floor has cooled down to equilibrium before loading. Failure to precool the room before loading is often the cause of unsatisfactory maintenance of temperature, slow cooling, and excessive shrinkage of the produce.

Temperature control

Air movement transfers heat from the produce to the coils by natural convectional circulation in a room with overhead grids (cooling pipes); by forced circulation in rooms cooled by forced or induced draft coolers; or by a combination of natural and forced convection. It follows that the nature of the packages and the method of stacking must allow the air to move readily through all parts of the stack for the produce to be cooled quickly and uniformly.

Spatial variation in produce temperature in a well designed and managed store should not exceed ±1°C of the nominated storage temperature. Several factors influence the spatial distribution of temperature in a store. The most important single requirement for uniform produce temperatures is uniform distribution of the cooling air over the entire surface area of the top of the stack. This applies equally to the distribution of air from fan-driven air circulation systems and to the even distribution of coils over the ceiling in natural circulation rooms. Also of importance is the uniformity of the air paths through the stow, since air always takes the path of least resistance. Ideally, there should be a continuous narrow air slot in the direction of air flow past at least two faces of every box or carton and past each side of every bulk bin, together with no large vertical gaps in the stack to allow the cool air to take a shorter path. The room should be well insulated to reduce heat leakage, and the coolers should have ample capacity to ensure a small difference between the temperature of the air and the coil surface.

Selection, sorting and handling of produce

It is desirable to sort and size-grade produce before storing it. Not all commodities are fit for storage; some have better keeping qualities than others, some are blemished but may be suitable for processing into fresh-cut or other products, and some of the harvested produce is unmarketable. Refrigerated storage is expensive, and it is not economic to have produce that is not fit for sale, or produce that would be better marketed immediately, occupying cool storage space. Sorting and sizing before storage, so that both the quantity and quality of goods in storage are known, assists orderly marketing.

Loading

If possible, warm produce should be cooled in a separate cool room from that used for storage. If only one room is available, the designed daily intake (commonly 10% of capacity) should not be exceeded. Otherwise, the life of the produce will be reduced and shrinkage promoted. Warm produce should be loosely stacked, and cooling can be improved with the aid of an auxiliary, portable fan placed in front of the stack, with the suction side to the produce, to draw air through it. More rapid cooling methods should be used for highly perishable produce (Chapter 4).

Stacking

Cool rooms should not be over-filled, as this results in variable temperatures and therefore a poor out-turn of produce. Packaged produce should be carefully stacked to give economy of space, adequate and uniform air circulation, and accessibility. This requirement is facilitated when the boxes of produce are unitised on pallets for mechanical handling. The following requirements are essential for rapid cooling and good temperature control when stacking any type of package:

1. Keep the stack 8 cm away from the outer walls and 10–12 cm away from any wall exposed to the sun. This will ensure that heat coming in through the walls will be carried away to the coils by air moving freely between the stack and the wall, without warming nearby produce.
2. Leave a clear air space of not less than 20 cm between overhead coil drip trays and the top of the stack. If unit coolers or other forced-air circulation systems are used, the clear space between the top of the stack (the load line) and the ceiling is generally not to be less than 25 cm and should not interrupt the flow of air. This ensures that a uniform layer of cold air blankets the whole stack. The full depth of the space in front of a forced draft cooler is kept clear for a distance of 2 m to allow it to function properly and to avoid freezing produce.
3. An air plenum of about 8 cm is required between the floor and the stack. When bins and pallets of boxes are used, the pallet bases provide the necessary air gap above the floor. Wherever possible, they are placed with the pallet bases parallel to the direction of airflow (i.e. running towards the forced draft cooler).

4. When boxes or carton without vents are used, leave small, vertical air paths within the stack, not less than 1 cm wide between adjacent packages, since heat exchange takes place mainly at the surface of the packages. A layer-reversed, open-chimney stacking pattern will provide the necessary vertical gaps between cartons (1 cm) and at the same time provide a stable stack. The advent of retail-ready packaging systems whereby the boxes or cartons occupy the whole area of the pallets has eliminated chimney stacking, and the packages fit tightly together. These packages are designed with vents in the side walls that are aligned to enable forced-air cooling. Provided that the produce is precooled to the desired temperature, there is usually enough air circulation around the stacks and through the vents to maintain produce temperature.
5. Bulk storage bins that may contain 200–500 kg of produce should have air gaps in the floor of 8–10 per cent of the base area. Rapid cooling of produce is possible in such bins in standard cool rooms. Bins of warm produce are first stacked only two-high overnight to allow quick removal of field heat from the produce. Next day they may be stacked to full height. Unless a high RH is maintained in the cool store, produce in bins that also have air gaps in the side may shrivel excessively. The sides of the bins can be lined to reduce shrivel, but cooling will be slow if the gaps in the base are covered. Around each column of bins – at least on the sides at right angles to the pallet-base bearers – vertical air gaps about 4 cm wide are left, as this allows free escape of the air rising by convection through the produce in the bin. Bins of produce may also be forced-air cooled.

Weight loss and shrinkage

Excessive shrinkage is caused by many factors. These include immaturity of the produce (lack of natural surface waxes), delay before storage, picking produce when it is hot and placing hot produce in the cool store, packing produce into dry wooden boxes or cartons, high storage temperatures including hot spots in the room, low RH due to insufficient insulation or undersized cooling coils, slow cooling, and excessive air circulation. Fast cooling, uniformly low temperatures, and high RH in the store are, therefore, necessary for low weight-loss. The extra cost of additional cooling and insulation, and an effective vapour barrier, are more than offset by reduced weight-loss and better produce condition after storage.

Weight loss during cooling may also be greatly reduced by wetting warm produce such as leafy vegetables with potable water, before putting it in the store. It is preferable to harvest produce early in the morning, when it is coolest, and to put it directly into the cool store. This reduces the load on the refrigeration plant and lowers costs. When it is necessary to harvest produce later in the day in hot, dry weather, it may be practicable to spray some types of produce with potable water and leave it in the open overnight to cool, by a combination of evaporative cooling and radiation cooling (if the night sky is clear), before placing the produce in the store the next morning.

Orderly marketing and over-storage

Over-storage is still a common fault in the cool storage of produce. It is sound marketing practice to commence selling long-keeping produce, such as apples and pears, from the cool store early and to continue regularly throughout the season. To achieve orderly marketing, produce in the cool store needs to be segregated according to the expected keeping quality and removed for sale accordingly; as a general rule, produce first in should be first out.

Over-storage of fruit may be minimised by transferring samples from each batch or line to room temperature at intervals during storage. At the first sign of deterioration of the sample, the whole line should be marketed without delay. A further important reason for regularly inspecting samples during storage is that some fruit, such as stone fruit, may seem to be in perfect condition in the cool store but may either develop physiological disorders or fail to ripen satisfactorily after removal to ambient temperature.

Sanitation

Cool rooms should be thoroughly cleaned at the end of each season and, if necessary, sterilised to reduce the risk of losses by mould attack. The walls and floor can be washed with a solution of sodium hypochlorite (chlorine). Mouldy or otherwise contaminated bins and boxes should be cleaned and sterilised with steam or a fungicide before reuse. Cleaning of boxes and bins has been made easier by gradually replacing wooden bins and boxes with plastic bins with smooth surfaces.

Grading machines are often an important source of mould contamination that leads to development of rots in storage. These machines should be cleaned and swabbed or sprayed with a fungicidal solution daily. The equipment should be regularly inspected for defects, and anything likely to cause produce injury should be remedied.

Installing ozone generators has been found to keep cool rooms free of surface moulds and to reduce the spread of fungal spores, but ozone is toxic to people. USA and Australian health regulations limit exposure of healthy individuals to a maximum of 0.3 ppm for not more than 15 minutes, repeated not more than four times per day. Individuals can be exposed repeatedly to 0.1 ppm during an 8-hour working day. When people can detect ozone by smell it has reached this concentration.

Refrigerated transport

Much produce is transported over long distances on land and sea under refrigeration. Refrigerated road or rail vehicles can be regarded as insulated boxes fitted with modular mechanical refrigeration units driven by electric or diesel motors. Refrigerated ships have a central refrigerating plant; the whole ('reefer ships') or only part of the carrying space on a vessel may be insulated and refrigerated.

Much refrigerated sea freight is now carried in containers of 30 or 60 m^3 capacity, which permit temperature control from door to door. Most refrigerated freight containers are 'integral containers', possessing their own electrical refrigeration unit and sometimes also a diesel-powered generator (Figure 7.7). Economy of space is a prime requirement in all transport; therefore, refrigerated transport vehicles and containers are designed for high-density stowage. They are not designed for rapid cooling, so successful refrigerated transport requires thorough precooling of the load. Respiratory heat is a significant proportion of the refrigeration load; consequently some air space must be provided between the packages during stowage of produce, unless the journey is short. Significant amounts of heat enter refrigerated road

Figure 7.7 Insulated shipping container fitted with a built-in refrigeration unit

Figure 7.8 To ensure effective air circulation in a refrigerated road vehicle, there must be an air delivery chute, ribs on the doors and walls, channels or pallets on the floor and a return bulkhead

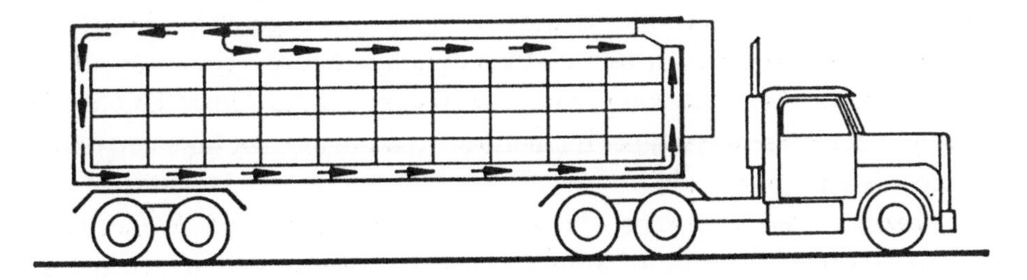

SOURCE A.K. Sharp, A.R. Irving and A.A. Beattie (1985) Transporting fresh produce in refrigerated trucks. *Agfact*, H1.4.3. NSW Department of Agriculture, Sydney.

transport vehicles from solar radiation, heat reflected from the road, and air leakage through the doors. To ensure even temperatures are maintained, it is necessary to provide good circulation of cooling air around the load (Figure 7.8). Rules covering precooling, stowage, and air circulation have been developed from research and commercial experience; maximum acceptable loading temperatures are commonly specified and should be closely policed.

Measuring the storage environment

The successful use of storage technology depends on being able to establish and maintain the desired environmental conditions in the storage chamber. This will certainly involve the measurement of temperature and possibly also RH, ethylene and the respiratory gases carbon dioxide and oxygen.

Temperature

A wide range of devices can be used to measure temperature. The simplest and most common is the liquid-in-glass thermometer. Mercury was commonly the liquid but ethyl alcohol that is coloured red is now the liquid of choice for safety reasons. Fluid-filled dial thermometers consist of a sensing bulb connected to a spiral tube filled with a fluid; these thermometers can take measurements at a distance of up to a few metres. They are often used in cool stores with an external indicating dial.

Remote temperature recording can also be achieved with thermocouples, which are two strips of different metals joined together at each end. When the junctions are at different temperatures, an electromotive force develops that is proportional to the difference in temperature. Resistance thermometers based on a thermistor operate similarly. They are robust, which allows their use for permanent distant reading installations; for example, in refrigerated ships. Temperature readings can be taken manually or recorded automatically.

Thermometers with a digital readout are now widely used. An electronic circuit detects the output of a thermocouple or thermistor, which is then converted to a temperature on a digital display. Digital thermometers can be hand-held and battery-operated or panel-mounted and mains-powered. A digital data-logger is an extension of the digital thermometer that measures temperature at preset intervals of time and stores the values in solid-state memory for later recall and analysis. Battery and mains-powered models are available. Data can be sent wirelessly to computer terminals or other forms of data logger. Moreover, the internet and/or the mobile phone network can be used to send temperature data over great distances in real time.

All temperature-measuring devices need to be calibrated; it should never be assumed that readouts are accurate. While it is impressive that temperatures on a digital thermometer can be displayed to a tenth of a degree, this can give operators the misleading impression that the data are correct. The accuracy of every device should be checked at 0°C with an ice-water mixture or against a pre-calibrated thermometer near the temperature at which it is to be used. Alternately, the device may be submitted to an approved testing laboratory for calibration. The accuracy of hand-held devices used in the field should be checked frequently.

The placement of temperature measuring devices in cool stores and similar structures is critically important. The nature of refrigeration systems means there is always a gradient in temperature within the chamber. The aim is to maintain the bulk of the produce at the recommended storage temperature, without freezing or over-cooling the coldest part. A thermostatic control device for forced-draft cooling is best placed in the air coming off the cooler, and adjusted to give the minimum acceptable temperature. The temperature of produce in a cool store should be measured in several positions due to the spatial variations in the chamber. Air temperatures just inside the door never give accurate readings of produce temperature.

Relative humidity (RH)

RH is not as easy to measure as temperature. In Chapter 5 we saw that there are various ways of defining the amount of water in air. Thus many methods have been devised to measure atmospheric water status, and no one hygrometer (or psychrometer) is really suitable for all purposes over the full range of RH and temperatures.

A wet and dry bulb hygrometer is the simplest and most widely used instrument for measuring RH. It consists of two thermometers, one of which, the dry bulb, measures the air temperature. The neighbouring wet bulb thermometer has a wet wick around the bulb, and evaporation of water from the wick into the atmosphere results in it being cooled. The temperature difference between the wet and dry thermometers can be translated to % of RH, water vapour pressure or dewpoint using tables prepared for this purpose. Precautions necessary for accurate reading include calibration of the thermometers, cleanliness of the wick, use of clean distilled or de-ionised water, ventilation of the bulbs with air moving at least 3 m/sec, and protection from external radiation sources. Wet and dry bulb instruments with thermistors instead of liquid-in-glass thermometers are also available; their advantages are small size and the possibility of automatic and remote operation.

In traditional hair hygrometers, the sensing element is a length of hair, or some other material capable of water sorption and desorption with a consequent change in length. The hairs are mechanically linked to a pointer on a scale that is generally printed on paper secured to a rotating drum that typically rotates once in 24 hours or in a week. Hair hygrometers have a number of weaknesses but are useful for monitoring slow variations in RH at almost steady temperatures, as in cool stores. They must be calibrated for each temperature range and have an accuracy of about 5% RH.

Electric hygrometers operate by recording variation in the resistance, capacitance or some other electrical parameter of a sensor with changing water sorption or desorption. Electrodes are attached to an insulating base containing a thin layer of an electrolyte that equilibrates with the surrounding air. The signals can be amplified and transmitted, which allows for remote usage. Capacitor sensors are more stable and reliable, except at high RH, and they must not become wet. In thin-film polymer hygrometers, the sensor consists of a thin solution film on a solid, making these instruments more robust and stable. Some can be washed with distilled water without losing calibration.

Chilled mirror sensors, although relatively expensive, have the distinct advantage of being relatively more reliable at high (>85%) RH. With the aid of appropriate electronics, a polished metal mirror is chilled until a thin film of condensation starts to form, changing its reflectance. Using algorithms similar to those used for wet and dry bulb thermometers, RH can be computed.

The calibration of most electronic RH measuring devices can be checked using saturated inorganic salt solutions. The precise headspace RHs over a variety of saturated inorganic salt solutions are published in reference handbooks.

Gas analysis

Carbon dioxide, oxygen and ethylene can be measured by both chemical and physical methods. However, the latter approach has been preferred because of relative speed, accuracy and sensitivity. In the laboratory, thermal conductivity gas chromatography is commonly used to analyse carbon dioxide and oxygen. An alternative method for rapid analyses that is widely used by industry, particularly for monitoring controlled atmosphere storage rooms, is to use an infra-red gas analyser for carbon dioxide, connected in series with a paramagnetic oxygen analyser. The sensors in these instruments are ventilated continuously with nitrogen, and gas samples are injected as for a gas chromatograph. The properties of the respiratory gases have been used to generate a wide range of portable hand-held analysers that are now commercially available to analyse for these gases. In addition, low-cost sensors based on chemical reactions have also been developed.

Ethylene levels are usually tested in a laboratory with flame ionization gas chromatography. Portable instruments using photo ionization detectors are now available and give quick results. Most portable analysers generate a digital image or printout of results, but it is important that these devices are calibrated with a certified gas mixture. This should be done even when the device contains an internal calibration system.

8 Physiological disorders

Physiological disorders involve plant tissue breakdown that is not directly caused either by pests and diseases or by mechanical damage – which includes tissue disruption upon ice crystal formation associated with freezing injury. Physiological disorders may develop in response to various pre- and postharvest conditions, including nutrient accumulation during organ development and low-temperature stress during storage, respectively. The visual appearance of some physiological disorders is shown in Plate 2.

Physiological disorders can be divided into five general categories: nutritional, temperature-related (low and high), respiratory, senescent and miscellaneous. Nutritional and low-temperature disorders are particularly problematic and are considered in more detail below. Sunburn on the shoulders of fruit, such as tomato and mango, is a common example of high-temperature injury incurred prior to harvest. Respiratory disorders are associated with low oxygen and/or high carbon dioxide concentrations in and/or around harvested produce in controlled atmosphere storage and modified atmosphere packaging. Black heart of potato is an example of low-oxygen injury and midrib browning of lettuce is an example of high–carbon dioxide injury. Senescence disorders, such as mealiness in apples, are generally associated with harvesting over-mature produce and/or over-storing produce. Other miscellaneous disorders tend to be cause- and/or product-specific in terms of the relatively unique symptoms expressed. For instance, exposure to ethylene causes russet spotting on the midrib of lettuce leaves and bitterness (isocoumarin accumulation) in carrots. Greening of potatoes exposed to light and rooting of onions exposed to high humidity may also be considered miscellaneous physiological disorders.

The cellular biochemical and biophysical mechanisms that give rise to physiological disorders in produce are extremely complex. Moreover, they often involve elusive interactions with the pre- and postharvest conditions (Figure 8.1). In diagnosing causes of physiological disorders or attempting to predict the likelihood of symptom expression in harvested produce, the following parameters may need to be taken into account: preharvest environment conditions (e.g. temperature, nutrition and water regimes), crop development factors (e.g. yield or crop load, position on the plant, and carbohydrate, water and/or nutrient partitioning), and postharvest environment conditions (e.g. temperature regime, gas atmosphere, and storage time).

Figure 8.1 A model for the development of postharvest disorders. Predetermined disorders develop late in maturation or during storage and are subject to modification by storage conditions; storage disorders are induced by the storage conditions, such as low temperature or high carbon dioxide, and preharvest conditions can modify their incidence and severity.

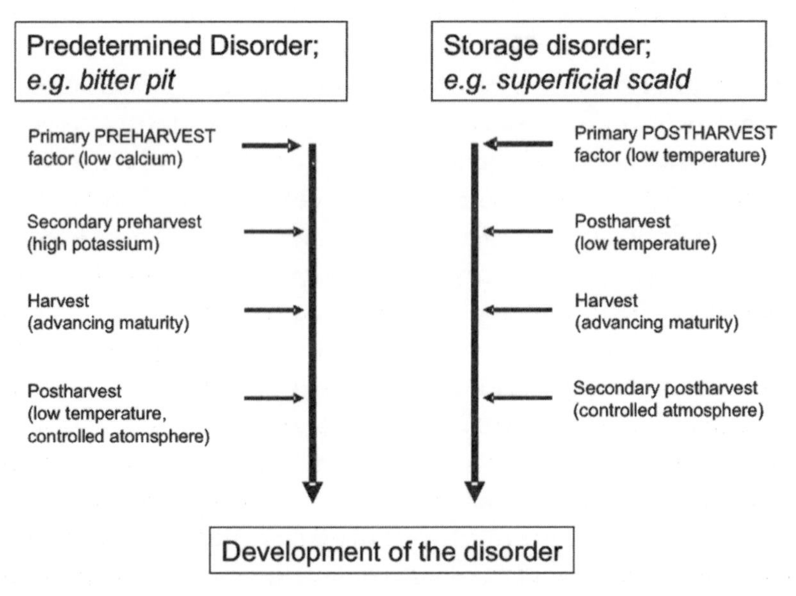

SOURCE I. Ferguson, R .Volz and A. Woolf (1999) Preharvest factors affecting physiological disorders of fruit. *Postharvest Biology and Technology* 15, pp. 255–62.

Mineral deficiency disorders

Disorders associated with deficiencies of specific minerals, and whose symptoms are sometimes expressed only after harvest in fruits, may be prevented by providing the specific mineral element either during growth or after harvest. For most mineral deficiency disorders, the actual role of the mineral in preventing the disorder has not been well established.

Calcium

Calcium (Ca) is associated with more postharvest-related deficiency disorders than any other mineral (Table 8.1). Some calcium deficiency disorders, such as blossom-end rot of tomatoes, can be eliminated by applying calcium salts as a preharvest spray. For others, such as bitter pit of apples, only partial control is obtained by preharvest sprays. Variability in the extent of control achieved is probably related to the amount of calcium taken up by the fruit. Postharvest dipping at sub-atmospheric pressures, which markedly increases the uptake of calcium, can result in total elimination of bitter pit. A substantial amount of the added calcium binds with pectic substances in the middle lamella and with cell membranes. Added calcium may possibly prevent some disorders by strengthening these structural components, without alleviating the original causes of the disorder. Strengthening cell components could prevent or delay the loss of sub-cellular compartmentation and the associated chemical and enzyme-mediated reactions that cause browning symptoms.

Table 8.1 Some calcium-related disorders of fruit and vegetables

Produce	Disorder
Apple	Bitter pit, lenticel blotch, cracking, internal breakdown, water core
Avocado	End spot
Bean	Hypocotyl necrosis
Brussels sprout	Internal browning
Chinese cabbage	Internal tipburn
Carrot	Cavity spot, cracking
Celery	Blackheart
Cherry	Cracking
Escarole (chicory)	Brownheart, tipburn
Lettuce	Tipburn
Mango	Soft nose
Parsnip	Cavity spot
Pear	Cork spot
Peppers	Blossom-end rot
Potato	Sprout failure, tipburn
Tomato	Blossom-end rot, blackseed, cracking
Watermelon	Blossom-end rot

Calcium deficiency may arise in one part of a harvested plant organ, resulting in symptoms in that specific region. Calcium has also been shown to affect the activity of many enzyme systems and metabolic sequences in plant tissues. Adding calcium to intact fruit or fruit slices generally suppresses respiration, but the response is concentration-dependent. The activities of isolated pectic enzymes have shown differential responses to calcium concentration. For example, the activity of pectinmethylesterase is initially increased by increasing concentrations of calcium but is inhibited at higher concentrations, while the large form of endopolygalacturonase is stimulated slightly by concentrations of calcium that inhibit the smaller forms of the same enzyme. The ability of calcium to regulate such enzyme systems has led to speculation that it may have a role in the normal fruit-ripening process. It is also possible that calcium prevents or delays the appearance of some physiological disorders by maintaining normal metabolism.

Other minerals

Boron (B) deficiency in apple leads to a condition known as internal cork. This condition is marked by pitting of the flesh and is often indistinguishable from bitter pit. The differences between the two disorders are that internal cork is prevented by applying boron sprays whereas bitter pit responds to calcium treatment, and that internal cork develops only on the tree, while bitter pit can develop after harvest.

The major mineral in plants is potassium (K), and both high and low levels of potassium have been associated with abnormal metabolism. High potassium has been associated with the development of bitter pit in apple, so both high potassium and low calcium are correlated with pit development. Potassium is associated with changes in tomato as it ripens and low potassium delays the development of full red colour by inhibiting lycopene biosynthesis.

Heavy metals, especially copper, act as catalysts for enzyme systems that lead to enzymic browning, such as browning of cut or damaged tissues that are exposed to air. The levels of these metals are important in processed fruit and vegetables, whether they are derived from the produce or from metal impurities acquired during processing.

Low-temperature disorders

Storing produce at low temperature is generally beneficial because the overall rate of metabolism (e.g. respiration, ethylene production) is reduced (Chapter 4). However, low storage temperatures do not suppress all cellular processes to the same extent. Some processes are especially sensitive to low temperature, and may cease completely below a critical temperature. Several cold-labile enzyme systems have been identified in plant tissues. Metabolic imbalance as a consequence of low temperature can lead to accumulation of reaction products and a shortage of reactants. If the imbalance becomes serious, essential substrates may not be produced and toxic products can accumulate. Consequently, cells will cease to operate properly and will lose their function and structure. Damaged cells often appear as discoloured areas (usually brown or black) (Plate 2 and Figure 8.2). Ethylene may be involved in low-temperature injury, since treatment with the ethylene binding site blocker 1-MCP (Chapter 6) can reduce discolouration symptoms associated with low-temperature disorders in some fruit (e.g. apple, pineapple). Metabolic disturbances occurring at sub-ambient temperature are generally divided into chilling injury and low temperature-associated physiological disorders, although both may involve similar cellular processes expressed on different (short and long, respectively) time frames.

Chilling injury

Chilling injury of specific produce is characterised by well-defined symptoms that are readily and reproducibly expressed in damaged tissues as a consequence of exposure to low temperature (Table 8.2). Chilling injury typically results from exposure of susceptible produce, especially that of tropical or sub-tropical origin, to temperatures below 10–15°C. However, the critical temperature at which chilling injury occurs varies among commodities. Chilling injury is completely different to freezing injury, which results when ice crystals form in plant tissues at temperatures below their freezing point. Both susceptibility to and symptoms of chilling injury are product- and even cultivar-specific. Moreover, the same commodity grown in different areas may behave differently in response to similar temperature conditions.

Table 8.2 Chilling injury symptoms of some fruits

Produce	Lowest safe storage temperature (°C)	Symptoms
Avocado	5–12*	Pitting, browning of pulp and vascular strands
Banana	12	Brown streaking on skin
Cucumber	7	Dark-coloured, water-soaked areas
Eggplant	7	Surface scald
Lemon	10	Pitting of flavedo, membrane staining, red blotches
Lime	7	Pitting
Mango	5–12	Dull skin, brown areas
Melon	7–10	Pitting, surface rots
Papaya	7–15	Pitting, water-soaked areas
Pineapple	6–15	Brown or black flesh
Tomato	10–12	Pitting, *Alternaria* rots

* A range of temperature indicates variability between cultivars in their susceptibility to chilling injury.

Skin pitting is a common chilling injury symptom that is due to collapse of cells beneath the surface. The pits are often discoloured. High rates of water loss from damaged areas may occur, which accentuates the extent of pitting. Browning or blackening of flesh tissues is another common feature of chilling injury (e.g. avocado; Figure 8.2). Chilling-induced browning in fruit typically appears first around the vascular (transport) strands. Browning can result from the action of the polyphenoloxidase (PPO) enzyme on phenolic compounds released from the vacuole during chilling, but this mechanism has not been proven in all cases. Fruit that has been picked immature may fail to ripen or ripen unevenly or slowly after chilling (e.g. tomato). De-greening of citrus fruit is slowed by even mild chilling. Water-soaking of leafy vegetables and some fruits (e.g. papaya) is also often observed. Symptoms of chilling injury normally occur while the produce is at low temperature. However, they sometimes only appear when the produce is removed to a higher temperature. Deterioration may then be quite rapid, often within a matter of hours.

Chilling injury causes the release of metabolites (e.g. amino acids, sugars) and mineral salts from cells. Leakage of metabolites and ions, together with degradation of cell membranes, provides substrates for growth of pathogenic organisms, especially fungi. Such pathogens are often present as latent infections or may contaminate produce during harvesting and postharvest storage, transport and marketing. Thus, increased incidence and severity of rots is another common symptom of chilling injury (Plate 2), particularly upon removal from low-temperature storage. Development

Figure 8.2 Chilling injury in avocado fruit appears as browning of the mesocarp due to the breakdown of cell compartmentalisation and the action of polyphenol oxidases to produce tannins (left). The fruit was stored in air for 21 days at 5°C and then ripened at 20°C for 5 days. The fruit on the right remained free of symptoms after ripening in air at 20°C following storage for 21 days at 5°C in sealed polyethylene bags that generated atmospheres typically comprising 7% carbon dioxide, 3% oxygen and 90% nitrogen.

SOURCE K.J. Scott and G.R. Chaplin (1978) Reduction of chilling injury in avocados stored in sealed polyethylene bags. *Journal of Tropical Agriculture (Trinidad)* 55, pp. 87–90. (Used with permission.)

of off-flavours or odours is another consequence of chilling injury.

The temperatures reported in Table 8.2 are limiting or critical temperatures below which symptoms of chilling injury will generally be observed. Symptoms of chilling injury, particularly those that become evident upon return to ambient conditions, may be more severe at temperatures further below the critical chilling temperature. Moreover, the symptoms can appear more quickly upon return to ambient conditions as a consequence of lower storage temperatures. The often complex relationship between storage life at various temperatures and sensitivity to chilling is illustrated in Figure 8.3.

The safest way to manage chilling injury is to determine the critical temperature for its development in a particular produce and then not expose the commodity to temperatures below that critical temperature. However, it has been found that exposure for a short period to chilling temperatures with subsequent storage at higher temperatures may prevent the development of injury. This conditioning process has been effective in managing black heart in pineapple, woolliness in peach, and flesh browning in plum. Modified atmosphere storage may also reduce chilling injury in some commodities. Finally, maintaining high RH both in storage at low temperature and after storage can minimise expression of chilling injury symptoms, particularly pitting (e.g. film-wrapped cucumbers).

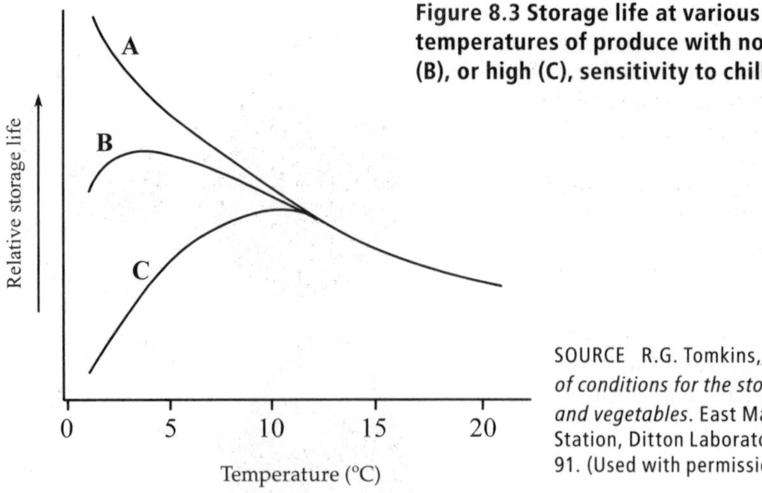

Figure 8.3 Storage life at various temperatures of produce with no (A), slight (B), or high (C), sensitivity to chilling injury

SOURCE R.G. Tomkins, *The choice of conditions for the storage of fruits and vegetables*. East Malling Research Station, Ditton Laboratory Memoir no. 91. (Used with permission.)

Mechanism of chilling injury

The critical temperature below which chilling injury occurs is an integrated genotypic by phenotypic characteristic of the particular organ. Highly chilling-sensitive organs, such as banana and pineapple fruit, have relatively high critical temperatures: 12°C or higher. It has even been suggested that the critical temperature may be greater than 20°C for some pineapple cultivars. Chilling-insensitive organs, such as apple and pear fruit, have much lower critical temperatures, around 0°C. Of course, low-temperature storage at below about –1°C is not possible for fresh fruit, vegetables or flowers because of freezing damage.

The cellular events of chilling injury can be separated into primary and secondary events. Primary events are virtually immediate and largely reversible (Figure 8.4). Secondary events are transiently reversible, but become irreversible; particularly with the onset of cell death and tissue necrosis. The main primary events in chilling injury are: low temperature–induced changes in the properties of cell membranes due to changes in the physical state of membrane lipids (membrane phase change), production of reactive oxygen species (e.g. hydrogen peroxide) that oxidise and thereby damage sub-cellular constituents including membranes, and dissociation of enzymes and other proteins into their structural sub-units such that enzyme activities are altered and structural proteins (e.g. tubulin) are disrupted. Data obtained using a range of techniques show changes in physical properties of extracted membrane lipids at temperatures in the range 7–15°C. These temperatures coincide with critical temperatures below

Figure 8.4 Time sequence of events leading to chilling injury

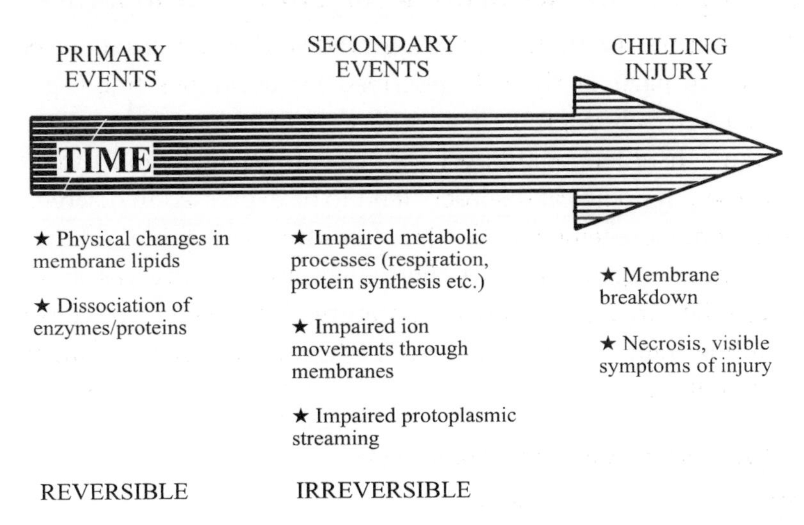

SOURCE G.R. Chaplin, personal communication.

which chilling-sensitive commodities express chilling injury symptoms. The physical changes in membrane lipids alter the properties of their parent membranes. Consequently, ion and metabolite movement across affected membranes, and also the activities of membrane-bound enzymes, are disrupted. The overall consequence of membrane disturbance is breakdown of sub-cellular compartmentation, which is readily measured as increased ion leakage from chill-injured tissues. Studies of both animal and plant systems reveal that enzymes involved in cellular metabolism may dissociate at low temperatures and some multimeric enzymes may split into component sub-units, with a consequent change in kinetic properties and perhaps loss of activity. Changes in the relative activities of enzymes lead to imbalanced metabolism and can ultimately result in cell death. Accumulation of toxic compounds (e.g. acetaldehyde) may accompany imbalanced metabolism. Also, structural proteins of the cell cytoskeleton (e.g. tubulin) dissociate in chilling-sensitive tissues at low temperatures.

Further research is required to fully elucidate the complex physical, biochemical and molecular mechanisms of chilling injury. If only a few proteins are involved in the synthesis of the key lipids and as cold-labile enzymes of metabolism, it may be possible to genetically manipulate these proteins to make plants less chilling-sensitive and thereby improve low-temperature storage prospects for sub-tropical and tropical fruits and vegetables.

Low temperature–associated physiological disorders

Low temperature–associated physiological disorders affect a range of fruit crops, but are particularly well described for deciduous tree (e.g. pome and stone) and sub-tropical citrus fruit crops (Tables 8.3 and 8.4). Low-temperature disorders also affect a range of vegetable and ornamental crops. These physiological disorders tend to be expressed in discrete areas of tissue. Some low-temperature disorders affect the skin of produce, but leave the underlying flesh intact. Others affect only certain areas of the flesh, or perhaps the core region. Low-temperature disorders may be considered to be chilling injuries that have developed slowly under low-temperature storage conditions.

Table 8.3 Some physiological disorders of apple

Disorder	Symptoms
Superficial scald	Slightly sunken skin discolouration, may affect whole fruit
Sunburn scald	Brown to black colour on areas damaged by sunlight during growth
Senescent breakdown	Brown, mealy flesh; occurs with over-mature, over-stored fruit
Low-temperature breakdown	Browning in cortex
Soft (or deep) scald	Soft, sunken, brown to black, sharply defined areas on the surface that extend a short distance into the flesh
Core flush (brown core)	Browning within core line
Water core	Translucent areas in flesh; may brown in storage
Brown heart	Sharply defined brown areas in flesh; may develop cavities

Table 8.4 Some physiological disorders of fruits other than apple

Produce	Disorder	Symptoms
Pear	Core breakdown	Brown, mushy core on over-stored fruit
	Neck breakdown, vascular breakdown	Brown to black discolouration of vascular tissue connecting stem to core
	Superficial scald	Grey to brown skin speckles; occurs early in storage
	Over-storage scald	Brown areas on skin in over-stored fruit
	Brown heart	Same as for apple
Grape	Storage scald	Brown skin discolouration of white grape varieties
Citrus	Storage spot	Brown, sunken spots on surfaces
	Cold scald	Superficial grey to brown patches
	Flavocellosis	Bleaching of rind; susceptible to fungal attack
	Stem-end browning	Browning of shrivelled areas around stem-end
Peach	Woolliness	Red to brown, dry areas in flesh
Plum	Cold storage	Brown, gelatinous areas on skin; flesh breakdown

For most physiological disorders, the metabolic events leading to symptoms manifesting are not fully understood. In fact, for many disorders, they are not understood at all. While the descriptions of some disorders might be considered non-scientific, this is nonetheless a critically important step towards optimising postharvest storage and handling practices. Cool store operators and shipping agents found a variety of browning conditions that developed on or in fruit held at low temperatures. These conditions were given descriptive names such as brown heart, as there was no other way of classifying the disorders. These descriptors remain the primary classification for low temperature–associated physiological disorders. Apple fruit has been studied more intensely than other commodities, which perhaps explains why the longest list of physiological disorders is for apples (Table 8.3).

Development of symptoms typically requires low-temperature storage, usually at less than 5°C. Each particular disorder is presumed to involve discrete aspects of metabolism, although this may not prove to be true when their biochemistry is fully elucidated. When more research effort is devoted to fruits other than apple, the list of physiological disorders will undoubtedly increase (Table 8.4). There is no reason to believe that apple should be more prone to physiological disorders than other commodities.

Early studies on low temperature–associated physiological disorders revealed that, although a particular variety may be susceptible to a certain disorder, not all fruit will develop the disorder. Susceptibility depends on various factors, including maturity at harvest, cultural practices, climate during the growing season, fruit size and harvest practices. The risk of a fruit developing a particular disorder can, therefore, be minimised by identifying susceptible fruit batches and not storing them for prolonged periods. However, the market often has a preference for types of fruit that are highly susceptible to a disorder. For example, the consumer often prefers large apples with intense red colouration, even though such fruits are susceptible to low-temperature breakdown. Thus, methods needed to be developed to successfully store susceptible produce and meet consumer requirements.

Various temperature-management programs have been developed to minimise the development of specific low temperature–related storage disorders. For some produce (e.g. persimmons, nectarines), visible symptoms of chilling injury may develop later and be less severe at temperatures closer to 0°C than at higher storage temperatures (e.g. 2–5°C).

The most commercially practical storage conditions for low-temperature injury–susceptible produce need to be determined for individual lines of produce. Low-temperature disorders can be considered a response to low-temperature stress. Plants have developed biochemical processes for coping with stress. These mechanisms include the expression of specific cold tolerance genes and of heat shock defence proteins. By deliberately invoking molecular, biochemical and enzymic mechanisms of stress-coping through specific conditioning treatments, susceptibility of harvested produce to low storage temperature stress may be ameliorated. For example, lowering the temperature in steps from 3°C down to 0°C in the first month of storage (i.e. step-down low-temperature conditioning) can minimise the development of low-temperature breakdown and soft scald in apple. Low-temperature breakdown of apple and stone fruits can also be reduced by periodically raising the temperature to around 20°C for a few days during the storage period (i.e. intermittent warming). To date, such methods have not been widely adopted in commercial practice because of the logistical problems of having a room full of uniform produce ready to treat at one time and the difficulty of rapidly changing the temperature of a room full of fruit. Another issue is that a transient increase in the storage temperature will shorten the storage life of any produce held in the same room that is not susceptible to the particular disorder (e.g. other varieties). Relatively brief periods of pre-storage exposure to intermediate low temperature (i.e. low-temperature conditioning), high-temperature stress (e.g. hot air, hot water dipping, hot water brushing), warm temperatures and high RH conditions (i.e. curing), or a nitrogen atmosphere (i.e. anoxia) have also proven beneficial in terms of reducing produce susceptibility to various low-temperature injuries. The preharvest temperature regime (i.e. periods of high or low temperature up to harvest) significantly influences postharvest susceptibility to low-temperature injury and response to conditioning treatments.

CA storage can also reduce the incidence of some disorders (e.g. core flush, Plate 2) and various forms of flesh breakdown in apple. Conversely, in some instances, the level of breakdown has been reported to actually increase under CA storage. This increase has been attributed to factors associated with controlled atmosphere storage other than the composition of the atmosphere. The enclosed room required for CA storage results in a high-humidity atmosphere, restricted ventilation rates, and accumulation of fruit volatiles in the atmosphere. These conditions are conducive to apple breakdown. Superficial scald (Plate 2) is another apple disorder that is exacerbated under CA storage by such conditions (see below). CA storage also exposes produce to high levels of carbon dioxide and low levels of oxygen for prolonged periods, which can cause some disorders. The critical level of carbon dioxide that induces brown heart of apple and pear varies among varieties, but may be as low as 1 per cent. Low-oxygen injury is characterised by alcoholic off-flavours from anaerobic metabolism, in addition to browning of the tissue.

The best way to prevent a disorder is to understand the metabolic processes that lead to the disorder and then implement measures to prevent that metabolic process from occurring. Chemical control is one approach to preventing disorders. However, storage disorders may also be minimised by physical and cultural treatments, and by breeding less susceptible cultivars.

Skin blemishes are a serious problem, as even quite small blemishes render fruit unacceptable in many markets. A low level of internal defects can be tolerated to a greater extent because the consumer buys on visual inspection of external appearance. Even upon consumption, the consumer may not become aware of a small amount of internal browning. The apple fruit skin disorders, bitter pit and superficial scald (Plate 2), have received considerable attention, and management measures have been developed for both disorders (Chapter 11).

Quite a lot is known about the metabolism of superficial scald. Early studies (before 1930) led to the hypothesis that it was caused by a toxic volatile organic compound that accumulated in the apple during cool storage. In the 1960s, Australian researchers isolated a 15-C sesquiterpene hydrocarbon, α-farnesene, from susceptible apple varieties. The oxidation products of α-farnesene are now claimed to lead to cell collapse and tissue browning. Superficial scald is controlled commercially by applying various synthetic antioxidants, such as diphenylamine and ethoxyquin to protect α-farnesene against oxidation. Chilling injury is usually regarded as developing through a different metabolic route to superficial scald. However, α-farnesene has been shown to accumulate in banana fruit during development of chilling symptoms, suggesting that there may be some metabolic commonality in the two disorders. More recent studies have shown that superficial scald can be controlled by ethanol vapour in the atmosphere surrounding Granny Smith apples. There may be resistance, based on religious and customs and excise grounds, to the commercial use of ethanol in this way. However, ethanol vapour is a natural substance that is metabolised, albeit in small quantities, by apple fruit.

If effective management methods are not available, the ultimate method of avoiding physiological disorder is to hold susceptible fruits at a temperature high enough to avoid the disorder being a problem. For commodities susceptible to low temperature–associated physiological disorders, this is usually 3–5°C, but is sometimes greater than 5°C. This strategy significantly negates the benefit of using low temperature to minimise metabolic rates, but slightly over-mature produce is easier to market than produce with an unsightly disorder.

9 Pathology

Postharvest losses of horticultural commodities by microorganisms from harvest to consumption can be rapid and severe, particularly in tropical areas where high temperatures and high humidity favour rapid microbial growth. Postharvest diseases affect a wide range of fruit and vegetables and the estimated losses are significant, especially in less developed countries due to lack of postharvest storage facilities. Furthermore, ethylene produced by rotting produce can cause premature ripening and senescence of other produce in the same storage and transport environment, and sound produce can be contaminated by rotting produce. Apart from actual losses due to wastage, further economic loss occurs if the market requirements necessitate sorting and repacking of partially contaminated consignments.

Many bacteria and fungi can cause postharvest spoilage of fruit and vegetables, but the major postharvest diseases are caused by species of the fungi *Alternaria, Aspergillus, Botrytis, Fusarium, Geotrichum, Monilinia, Penicillium, Rhizopus* and *Sclerotinia* and of the bacteria *Erwinia* and *Pseudomonas* (Table 9.1 and Plate 3). *Botrytis* is the most common and commercially important pathogen of cut flowers. Most of these organisms are weak pathogens in that they can only invade damaged or senescing produce. A few, such as *Colletotrichum*, are able to penetrate the skin of healthy produce. Often the relationship between the host fruit, vegetable or ornamental and the pathogen is reasonably specific. For example, *Penicillium digitatum* rots only citrus fruit and *P. expansum* rots apple and pear fruit, but not citrus. Complete loss of the commodity occurs when one or a few pathogens invade and break down the tissues. Initial breakdown is often rapidly followed by invasion by a broad spectrum of weak pathogens and saprophytes, which magnify the damage caused by the primary pathogens. The appearance of many commodities may be marred by surface lesions caused by pathogenic organisms, without the internal tissues being affected.

Table 9.1 Examples of major postharvest diseases of fresh fruits and vegetables

Crop	Disease	Pathogens
Apple, pear	Grey mould	*Botrytis cinerea*
	Blue mould	*Penicillium expansum*
Banana	Crown rot	*Colletotrichum musae, Fusarium roseum, Verticillium theobromae, Ceratocystis paradoxa*
	Anthracnose	*Colletotrichum musae*
Citrus fruit	Green mould	*Penicillium digitatum*
	Blue mould	*Penicillium italicum*
	Sour rot	*Geotrichum candidum*
Grape, strawberry, leafy vegetables	Grey mould	*Botrytis cinerea*
Papaya, mango	Anthracnose	*Colletotrichum gloeosporioides*
Peach, cherry	Brown rot	*Monilinia fructicola*
Peach, cherry, strawberry	Rhizopus rot	*Rhizopus stolonifer*
Pineapple	Black rot	*Ceratocystis paradoxa*
Potato, leafy vegetables	Bacterial soft rot	*Erwinia carotovora*
	Dry rot	*Fusarium* spp.
Sweet potato	Black rot	*Ceratocystis fimbriata*
Leafy vegetables, carrot	Watery soft rot	*Sclerotinia sclerotiorum*

The infection process

Fruit and vegetables are rotted by microorganisms that infect the produce either while it is attached to the plant or during harvesting and subsequent handling and marketing operations. The infection process is greatly aided by mechanical injury to the skin of produce – such as insect punctures, cut stems, fingernail scratches, abrasions and rough handling. Furthermore, the developmental stage and stress condition of the produce, environmental conditions (e.g. temperature) and natural host defences (e.g. formation of periderm; see later in this chapter) significantly affect the infection process

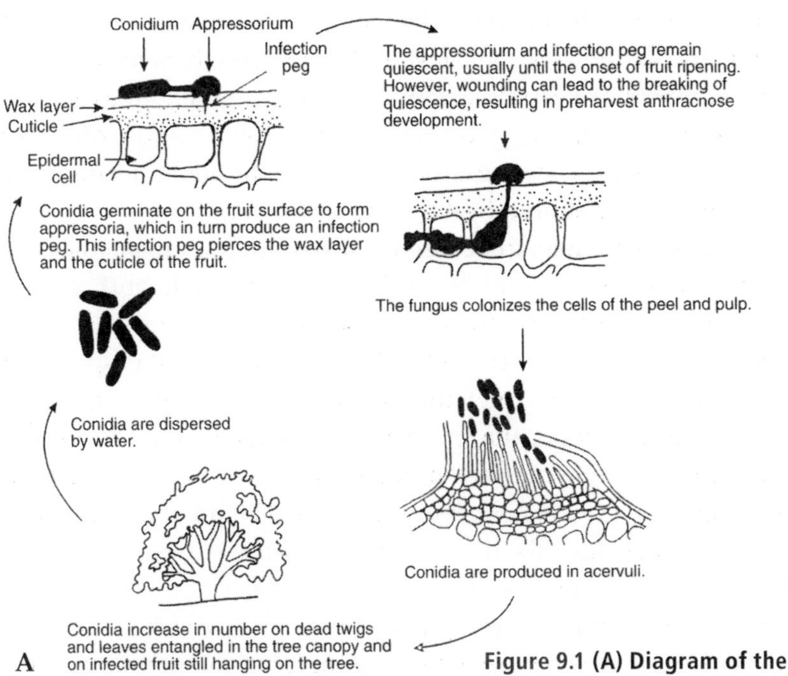

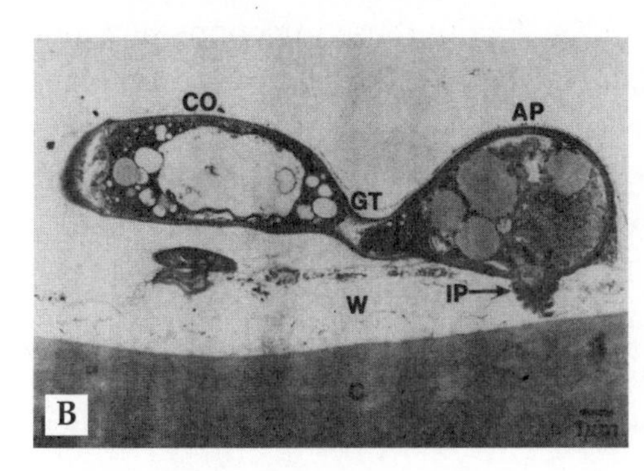

Figure 9.1 (A) Diagram of the characteristic anthracnose disease cycle caused by the plant pathogen *Colletotrichum gloeosporioides* and (B) a transmission electron micrograph of a *C. gloeosporioides* spore germinating on the skin of an avocado fruit (CO = conidium, GT = germ tube, AP = appressorium, IP = infection peg, W = wax layer, and C= cuticle)

SOURCE Images provided by Tony Cooke and Dr Lindy Coates, Department of Primary Industries and Fisheries, Queensland.

and the subsequent invasion process. It is important to understand the infection and invasion process in order that suitable treatment strategies can be implemented to prevent or manage the infection (Figure 9.1).

Preharvest infection

Preharvest infection of fruit and vegetables may occur through direct penetration of the skin, infection through natural openings on the produce, infection spreading from the parent plant, and infection through damage. Some pathogenic fungi are able to initiate an infection on healthy floral parts and on sound developing fruit. The infection is then often arrested and remains quiescent until after harvest, when the natural resistance of the host decreases and/or conditions become favourable for growth. *Botrytis cinerea* can infect healthy petals of certain cut flowers (e.g. carnation, freesia, rose) prior to harvest and is typically arrested by the hypersensitive response until flowers start to senescence after harvest. These limited infections are evident as unsightly discoloured flecks or specks.

Organisms that form latent infections are also important in postharvest diseases of many tropical and subtropical fruits – for example, anthracnose disease of mango and papaya, crown rot disease of banana and stem-end rot disease of citrus. An example of this process is the spores of *Colletotrichum* that germinate in moisture on the surface of the fruit; within several hours of germination, the tip of the germ tube swells to form a hard-walled structure known as an appressorium. An infection peg from the appressorium may or may not penetrate the skin before the infection is arrested. At some later time, the infection will become active and the host tissue will be extensively invaded by the fungal hyphae.

Other pathogenic and/or saprophytic fungi and bacteria can gain access to developing fruit and vegetables through natural openings, such as stomates, lenticels and growth cracks. Again, these infections may not develop until the host becomes less resistant to the invading organism – for example, when the fruit ripens and senesces. An example of this infection mechanism is the penetration of apple lenticels before harvest by spores of *Phlyctaena vagabunda*, which then manifest in storage as rots around the lenticels. Most sound fruit and vegetables can suppress the growth of pathogenic organisms for a considerable time due to preformed (e.g. cuticle) and induced (e.g. lignification) natural physical barriers (e.g. epidermis, periderm) and chemical defence compounds (e.g. antifungal phenols, pathogenic proteins).

Postharvest infection

As noted above, many of the postharvest pathogens that cause considerable loss of horticultural produce are unable to penetrate intact skin of produce, but readily invade via any mechanical damage in the skin. Even microscopic damage is sufficient for the spores and other infectious structures of pathogens present on the produce and in the packing house to gain access to the produce. The cut stem is a common point of entry for microorganisms. Thus, stem-end rot diseases are important in postharvest wastage of many fruits and vegetables.

Dying and dead floral remnants can be a ready source of postharvest infection. In the case of grape and Geraldton wax flower, *Botrytis cinerea* that infected senescing or necrotic anther tissues readily invades berries and other flower parts, respectively, after harvest. As mentioned earlier, postharvest infection can also occur through direct penetration of the skin by strong pathogens like *Sclerotinia* and *Colletotrichum*.

Factors affecting development of infection

Probably the most important factor affecting development of postharvest wastage caused by pathogens is the surrounding environment. High temperature, high humidity and free water favour development of postharvest decay. Chilling injury also predisposes tropical and subtropical produce to postharvest decay. In contrast, low temperature, low oxygen and high carbon dioxide levels and the correct humidity can restrict the rate of postharvest decay by either retarding the rate of ripening or senescence of produce, depressing growth of the pathogen, or both.

Many other factors affect the rate of development of an infection in fruit and vegetables. The host tissue, particularly the pH of the tissue, acts as a selective medium. Fruit generally has a pH below 4.5 and is largely attacked and rotted by fungi. Many vegetables have a pH above 4.5 and, consequently, bacterial rots are more common. Ripening fruit is more susceptible to wastage than immature fruit, so treatments such as low temperature that slow down the rate of ripening will also retard growth of decay organisms. The underground storage organs, such as potato, cassava, yam, and sweet potato, are capable of forming layers of specialised cells (wound-periderm) at the site of injury, thus restricting the development of postharvest decay. During commercial handling of potato, periderm formation is promoted by 10–14 days of storage at 7–15°C and 95% RH, a process known as curing. A type of curing process (possibly by desiccation) has been shown to reduce the wastage of orange by *P. digitatum*. When the fruit is held at high temperature (30°C) and humidity (90%) for several days, the peel becomes less turgid and lignin is synthesised in the injured flavedo tissue.

Control of postharvest diseases

An integrated approach to disease management encompassing both preharvest and postharvest activities should be considered for optimum effectiveness.

Preharvest disease control

In most instances, control of postharvest diseases can commence before harvest, in the field or orchard. Wherever possible, sources of infection should be eliminated, and sprays for the control or eradication of the causal organisms applied. Preharvest sprays are generally not as effective as postharvest applications of the chemical directly to the commodity, although some systemic fungicides have shown good control of latent infections, such as lenticel rot of apple and brown rot of peach, when applied prior to harvest. With some of the newer, more specific fungicides, organism resistance has occurred. For instance, *Penicillium* species rapidly developed resistance to the benzimidazole group of fungicides, which suggests that preharvest sprays of these fungicides would be very unwise, because of the opportunity for selection and growth of resistant strains, particularly if the same fungicide was also relied upon for postharvest control of *Penicillium.*

Careful handling during harvesting can minimise mechanical damage and thus reduce subsequent wastage due to microbial attack. The weather at harvest should be dry and cool to avoid further contamination and infection. It is unwise to harvest some fruits after rain or heavy dew, such as citrus, whose peel is turgid under those conditions and more easily damaged. Other management practices such as cooling the crop as soon as possible after harvest to minimise microbial growth and removing produce showing signs of infection to avoid the spread of the disease will help to control postharvest diseases.

Postharvest disease control

Postharvest disease control methods can be classified under three broad categories: chemical, physical and biological treatments. The effectiveness of a treatment depends on three main factors: ability to reach the pathogen, level and sensitivity of the infection, and sensitivity of the host produce.

The time of infection and the extent of development of the infection are critical in respect to whether it can be controlled. For example, *Penicillium* and *Rhizopus* invade wounds during harvest and subsequent handling operations and are relatively easily controlled by fungicide application to the surface of the commodity. In contrast, *Botrytis* infects strawberry fruit in the field some weeks before harvest or even at the time of flowering and is therefore relatively difficult to control. It is recommended that fungicides be applied within 24 hours of harvest so that infections can be controlled before they become established.

Chemical treatments

Chemical control of postharvest diseases has become an integral part of the handling and successful marketing of fruit only during the past 30 years, particularly in the development of the world trade in citrus, banana and grapes. The reduction in losses depends on the marketing strategy for the commodity and the type of infection. For citrus, which has a relatively long postharvest life, the aim of the treatment is to prevent primary infection and also sporulation so that nearby fruits are not contaminated. The strawberry has a short postharvest life, and treatment is aimed at preventing the spread of the field-induced infection of grey mould (*Botrytis*). In other words, the treatment has to match the subsequent marketing of the commodity. It is neither necessary nor desirable to treat a short-life commodity with a fungicide that has a long residual activity.

The success of a chemical treatment depends on several factors: initial spore load, depth of the infection within the host tissues, growth rate of the infection, temperature and humidity, and depth to which the chemical can penetrate the host tissues. Moreover, the applied chemical must not be phytotoxic (i.e. must not injure the host tissues) and must be permitted for use by the local food regulations.

Over the years, a wide range of chemicals has been used for the control of postharvest losses in fruit, particularly citrus, banana, grapes and strawberry. The range of chemicals include: alkaline inorganic salts such as sodium tetraborate (borax); ammonia and aliphatic amines such as *sec*-butylamine; aromatic amines such as dicloran; benzimidazoles such as benomyl; triazoles such as imazalil and prochloraz; guanidines such as guazatine; hydrocarbons such as biphenyl; oxidising substances such as iodine; organic acids and aldehydes such as sorbic acid and formaldehyde, respectively; phenols such as sodium ortho-phenylphenate (SOPP); and salicylanilide and sulphur as inorganic compounds such as sulphur dust and sulphur dioxide gas and as organic compounds such as captan. Novel fungicides being developed and evaluated for controlling postharvest diseases include pyrimethanil for control of *Botrytis*; and strobilurins, such as azoxystrobin, to control a range of postharvest fungal pathogens.

These chemicals are generally fungistatic in action rather than fungicidal. That is, they inhibit spore germination or reduce the rate of germination and growth rather than killing the organism, and they must come into direct contact with the organism to be effective. A few chemicals, such as chlorine, iodine and sulphur dioxide, are true fungicides. Chlorine and iodine are commonly added to wash-water to kill bacteria and fungi, and sulphur dioxide is lethal to *Botrytis* on grapes. Depending upon their chemical and physical properties and formulation, various fungicides may be impregnated into wraps or box liners, or applied as fumigants, solutions, suspensions, or in wax.

Suitable selection and application of chemical fungicides can greatly improve the outcomes of fruit and vegetable storage and transportation. For example, black-end, finger-stalk or crown rot of banana, and brown rot in peach, are effectively controlled by the benzimidazoles; grey mould in table grapes by the slow introduction of sulphur dioxide from an in-package generator or by room fumigation with sulphur dioxide; and *Rhizopus* rot in stone fruits by dicloran. In contrast, control of *Geotrichum*, *Alternaria* and the soft rots is still unsatisfactory, although SOPP does give some control of *Geotrichum* when applied 4–24 hours after infection.

The continuing evolution of modern postharvest chemical control can be illustrated with citrus fruit. Green (*Penicillium digitatum*) and blue (*P. italicum*) moulds are the major postharvest diseases of citrus, with green mould being more prevalent in humid areas. The stem-end rots are also significant causes of postharvest losses in more humid climates. The advent of the benzimidazoles, such as thiabendazole (TBZ) and benomyl, in the 1970s rapidly superseded earlier chemical treatments because of their wide spectrum of anti-fungal activity at extremely low concentrations, although they are inactive against *Rhizopus, Alternaria, Geotrichum* and soft rot bacteria. The other advantage of the benzimidazoles is that they are systemic and therefore have substantial residual activity. For example, benomyl was able to provide excellent wastage control in lemons stored for six months at 13°C. Because of the ease with which these compounds may be handled, packing houses rapidly adopted the benzimidazoles, but resistant strains of *Penicillium* have developed, giving rise to a need for alternative treatments. Benomyl is no longer registered as a postharvest treatment in various parts of the world, although other benzimidazoles (e.g. carbendazim and TBZ) are still registered at present for postharvest treatment of various crops. The need for alternative chemicals led to the introduction of *sec*-butylamine, guazatine, and the triazoles, particularly imazalil and prochloraz, which are also very effective against a wide range of fungi, including some that are now resistant to the benzimidazoles. However, all chemicals must be considered to have a limited commercial life, and resistant strains to the newer chemicals as well have now appeared.

If chemical fungicides are to remain in the arsenal of microbial control, a continuing need exists to develop new compounds. The ideal postharvest fungicide should be water soluble, have broad-spectrum activity, not be phytotoxic, be safe to use (i.e. leave no residues toxic to consumers), not affect palatability, remain active over a long period, leave no visible residues, and be cheap – that is, the compound should either be inexpensive or effective at low concentration. Few, if any, of the present fungicides meet all of these requirements.

An added consideration is the increasingly stringent process that new fungicides must navigate before they are approved for use by the appropriate government authorities in each country. This procedure is in place to ensure new chemicals are not toxic at the concentrations used, to set maximum residue levels (MRLs) for approved fungicides, and to empower government inspection services to check that the products are only being used for the specific approved purpose. Despite this protocol, there is growing public concern about the use of synthetic chemicals, and pressure is building for the horticultural and agricultural industries to use alternative, non-chemical means of disease control. Physical treatments and biological controls are perceived as more desirable by consumers; they are seen as more natural and they avoid the residues arising from chemicals fungicides.

The use of endogenous plant metabolites to control microbial growth could be a method of chemical treatment perceived by the consumer as being more acceptable as these chemicals are generally recognised as safe (GRAS). The organic volatile compounds produced by all fruit and vegetables are being actively examined, and in some cases, re-examined, for anti-microbial activity. The essential oils, mainly mono- and sesquiterpenes, have long been known to have anti-microbial activity, a property that has assisted their longstanding use in human therapy. Many of the flavour volatiles, especially aldehydes such as hexanal and alcohols such as ethanol, inhibit decay on a range of produce. Their volatility allows them to be readily used as vapours released into the atmosphere around produce during storage or transport. A number of issues still need to be resolved before commercialisation of these compounds can occur. Not the least is that most need to be applied at much greater concentrations than the levels at which they occur naturally in produce. Commercial quantities would most likely be obtained through chemical synthesis rather than extraction from plant material. This raises the question of when something ceases to be natural and becomes an applied chemical in the eyes of the regulators and consumers.

Physical treatments

Physical treatments involve handling procedures that minimise postharvest injury and hence the entry of microorganisms into the commodity, and also creating an external environment unfavourable to the growth of microorganisms.

The importance of careful handling in minimising damage was been stressed earlier, and it is mentioned again because it is low cost and can be implemented in even the most basic postharvest system, but results in dramatic benefits in marketing through reduced microbial wastage. Associated with careful handling is a good sanitation program in the postharvest environment that minimises the background level of spores and hence reduces their contact with produce. An important benefit of packaging is the protection it provides against damage during handling and transport and against cross contamination (see Chapter 12).

The use of low temperature during handling and storage is the most important physical method of postharvest wastage control, and all other methods can be considered as supplements to low temperature. While the susceptibility of most tropical and subtropical produce to chilling injury limits the benefits to be gained from low temperatures, holding such produce at, say, 15°C is more beneficial than leaving it at ambient temperatures, which can be greater than 30°C.

Heat treatments, using either moist hot air or hot water dips, are used commercially to control postharvest wastage in papaya, mango, stone fruits and cantaloupe. Advantages of hot water dipping are that it can control surface infections as well as infections that have penetrated the skin but it leaves no chemical residues on the produce. The beneficial effect is at least partly due to the enhanced formation of lignin and related compounds, which prevent invasion by germinating mould spores. The absence of chemical residues demands that recontamination of the produce by microorganisms be prevented by strict hygiene and possibly application of a fungicide, although at lower levels than those required without a hot water dip. Hot water dips must be precisely administered, as the range of temperature (commonly 50–55°C) necessary to control wastage approaches temperatures that damage produce. A relatively new technology, hot water brushing, has been developed in Israel and has proven highly effective as a disease control treatment for a range of amenable fruit crops, including peppers and citrus.

UV radiation can be directly used against postharvest pathogens by treating water used in packing houses, but the commercial viability of such treatment is still questioned. Irradiation of fruit, particularly citrus, with UV light at 254 nm has been found to effectively inhibit mould growth. The use of UV light was further enhanced by subsequent exposure to a hot water dip. It appears to stimulate the fruit to increase its production of lignin in the surface layers and of endogenous phytoalexins, such as scoparone, that are the natural defence system of the fruit against microbial invasion. The stimulation of phytoalexins following exposure to UV light has now been recorded for a number of fruits and vegetables.

Other important factors in minimising microbial growth, including the use of modified atmospheres with elevated carbon dioxide, reduced oxygen and reduced ethylene levels, and better control over humidity conditions around produce have been mentioned earlier. Ionising radiation can inhibit microbial growth but may adversely affect fruit and vegetables. The technology has found limited application around the world due to resistance from consumers and politicians.

Biological treatments

Before the advent of the synthetic chemical industry, biological control methods were often the main tool used to prevent degradation of food crops. However, it is a lengthy process to discover a potential biological control agent for use on fresh produce, prove its safety and effectiveness on the target crop, and confirm that it will be able to survive in the new environment without becoming a problem and having undesirable side effects. Nonetheless, with the likely eventual loss of effectiveness of all chemical treatments due to development of species resistance, and with consumer dislike of synthetic chemicals applied to foods, interest in biological control methods for postharvest diseases has been increasing in recent years.

While many biological agents (Table 9.2), including fungi, yeasts and bacteria, may potentially be useful in controlling postharvest diseases, identifying and commercially evaluating these is at a relatively primary stage. For biological control to be effective in the postharvest environment, the storage conditions must allow the biological agent to remain viable, and an efficient application method is required to fully cover produce surfaces. Needless to say, the treatment needs to pay for itself by generating a greater economic return. Biological control agents can suppress pathogens by a range of mechanisms or combinations of mechanisms. Direct attack, competition for space and nutrients, and eliciting heightened host defence mechanisms are generally considered acceptable mechanisms. A biological agent may also produce chemicals that are toxic to the disease organism, i.e. by synthesising natural antibiotics. This mechanism is generally not considered acceptable for food crops, but could be accepted for ornamental crops.

Table 9.2 Potential postharvest biological control agents

Commodity	Disease	Biological Agent
Apple	Blue mould	*Pseudomonas syringae* *Pseudomonas cepacia* *Cryptococcus* spp. *Candida oleophila*
	Grey mould	*Pseudomonas cepacia* *Cryptococcus laurentii* *Acremonium breve*
Pear	Blue mould/grey mould	*Pseudomonas cepacia*
	Grey mould	*Pseudomonas gladioli*
	Mucor rot	*Cryptococcus laurentii* *Cryptococcus flavus*
Stone fruit	Brown rot	*Bacillus subtilis*
Citrus	Sour rot	*Bacillus subtilis* *Trichoderma* spp.
	Green mould, blue mould	*Candida oleophila*
Grape	Grey mould	*Trichoderma harzianum* *Pichia guilliermondi*
Strawberry	Grey mould	*Trichoderma harzianum*
Pineapple	Penicillium rot	Attenuated strains of *Penicillium* spp.
Potato	Soft rot	*Pseudomonas putida*
Tomato	Grey mould, *Alternaria* rot	*Pichia guilliermondi*

As an example of the promise of biological control, *Rhizopus stolonifer*, the cause of transit rot in peach, has been controlled with high concentrations of the bacterium *Enterobacter cloacae* (10^{12} bacteria per mL). Also, the bacterium *Bacillus subtilis* has been shown to control brown rot of peach when sprayed as a suspension of 10^9 live bacteria per mL. *B. subtilis* has also been found active against citrus green mould, sour rot and *Alternaria* centre rot. In these cases, the controlling agents are thought to be antifungal substances produced by the bacterium that prevent mould development. Brown rot of peach has also been reduced by two as yet unidentified antibiotic substances isolated from the growth media of the fungus *Penicillium frequentans*.

Antagonism between organisms is the most promising area of biological control. It involves growth of a non-pathogenic organism preventing growth of a pathogenic organism. The antagonistic microorganism may be added from an external source, or indigenous microflora on the surface of the host may be stimulated in such a way that the incoming pathogen cannot develop on the plant surface. An example is the yeast *Debaryomyces* (now called *Pichia*), which appeared to reduce green mould development by competing for space and nutrients in an injury site on the fruit rind of citrus, thus inhibiting mould development. It does not produce an antibiotic, which has advantages as such compounds may have toxic effects on consumers.

While antifungal substances or the organisms themselves may occur naturally, they will have to be rigorously tested for human toxicity and carcinogenicity in the same manner as currently used fungicides. Biological control methods may have additional problems, such as inducing allergic responses in humans. A number of microbial biopesticides have been cleared for postharvest use. Aspire™ is available for use against *Botrytis* and *Penicillium*, and Bio-Save 110 and 1000™ for use against postharvest pathogens on apple, citrus and pear. Selected commercial strains of *Trichoderma harzianum* have demonstrated potential for control of grey mould (*Botrytis*) on ornamentals.

10 Evaluation and management of quality

The term 'quality' defies complete and objective definition. For each consumer of horticultural produce, quality is a highly subjective judgement related to learned criteria. An example is orange fruit afflicted by a mite that causes the skin to go bronze – some individuals 'in the know' register the bronzed appearance (quality criterion) as a promise of a very sweet orange, while most register it as the appearance of a fruit to be avoided.

Quality criteria for fruits, vegetables and ornamentals vary between commodities. Furthermore, for any particular commodity, the definition and associated criteria also depend on the perspective of the recipient. For example: what is a good quality pear?

- To the producer, a good quality pear is one that secures a maximum price in the market at a particular time of the season. Judgement can be exercised to balance producing a high quality product by intensive care but with a low yield against a poorer quality product by less intensive care but with a higher yield. For example, a higher total return may be gained for the higher price of early picked pears, despite the quality not being ideal for the consumer and the crop yield being lower than if picked later.
- To the shipper, a good quality pear is hard and green. It can be transferred from orchard to market without bruising or ripening.
- To the canner, good quality is a ripe but firm pear. The consumer requires a canned pear to be soft, but the pear needs to be firm enough to retain its shape undamaged during processing and subsequent market handling.

- To the consumer of the fresh fruit, a good quality pear is soft and ripe. It must melt in the mouth and be juicy. Skin colour is also often important, whereas to the canner this is immaterial as the skin is removed prior to processing.

Quality may, therefore, be defined in terms of end use or 'fitness for purpose'. In this context, produce quality requirements commonly refer to market, storage, transport, eating and/or processing quality. The marketing of fresh fruit and vegetables and of ornamentals is aimed eventually at the consumer, for whom tradition (i.e. learned criteria) plays a major role in determining acceptability. Consumer purchasing habits are typically conservative, so inducement is often required to get people to try something new, for example a pear with a red skin.

Because of markedly different quality considerations between ornamentals and fruit and vegetables grown for eating, postproduction quality evaluation for ornamentals is considered under a separate heading, later in this chapter.

Quality criteria

Quality criteria can be divided into external and internal factors. In terms of selling, external criteria might be considered of paramount importance. Accordingly, measures are often taken to try to 'improve' external quality. Examples of such measures include waxing of apples, degreening of oranges, reddish lighting to enhance the appearance of red apples, and orange-coloured mesh bags to reinforce the colour of oranges. However, if the consumer is disappointed by poor internal quality, there will be a reduced number of repeat purchases. For instance, consumers are often disappointed by the poor organoleptic properties of early season fruit (e.g. immature nectarines that fail to ripen properly) or out of season fruit (e.g. mealy apples that have been over-stored). Some important quality criteria for consumers are: appearance, including size, colour and shape; surface and internal defects, mouthfeel or texture; flavour; and nutritional value.

Appearance

People 'buy with their eyes' and learn from experience to associate desirable qualities with a certain external appearance. A rapid visual assessment can be made on the basis of size, shape, colour, condition (such as freshness), and/or the presence of surface defects or blemishes.

Size can be easily measured by circumference, diameter, length, width, weight or volume. Many fruit are graded according to size, often by diameter measurement, with similar sizes of fruit being packed together to facilitate marketing and retail sales (Plate 4). For example, certain size standards are adhered to for export apples based on fruit diameter or the associated number per package. These standards are specific to a particular package and also depend on the export destination. Particular packing arrays and individual wrapping of fruit, especially the top layer, may also be mandatory or recommended. The packing array will determine the number of fruits per package for some commodities. An example of a commodity graded by length and diameter is the carrot. For many other commodities, weight is the standard determinant.

Shape is a criterion that often distinguishes particular cultivars of fruit. Characteristic shapes are usually demanded by the consumer, who will often reject a commodity that lacks the characteristic shape. For example, attempts to market a straight banana were unsuccessful, apparently because this shape was considered abnormal. The amount of bend in banana is stipulated in the EU market, so consumers there never know that a banana can be straighter. Shape is especially important in apple cultivars, a premium price being obtained for fruit with a well-developed shape, characteristic of the particular cultivar. Misshapen fruit and vegetables are poorly accepted and usually bring a lower price. Shape is often a problem in breeding programs. Although a superior eating or storing product may be obtained by breeding, if its shape is unusual it will be less readily accepted in the market and extensive consumer re-education through advertising will be required.

One of the distinguishing features of fruit and vegetables is that they are the only major group of natural foods with a variety of bright colours. They are often used merely to brighten up the presentation of foods. Parsley contains relatively high levels of ascorbic acid, carotene, thiamin, riboflavin, iron and calcium compared to fruit and other vegetables, yet in Western society it is used almost exclusively to add colour and flavour to meat and fish dishes. Consumers correlate colour changes in ripening fruit with sweetening (the conversion of starch to sugar) and the development of other desirable attributes, so the correct skin colour is often all that is required for a decision to purchase. However, such subjective assessments may be misleading. For example, if bananas are ripened at higher than optimum temperatures, full loss of green colour does not occur, and consumers, particularly those from temperate climates, would show buyer resistance even though the flesh may be adequately ripe. Standardised colour charts are used in the visual assessment of ripeness in many fruits, for example, tomato, pear, apple and banana (Plates 5 and 6).

Defects

Normal appearance is extremely important in the marketplace. Consumers have a firm idea about what constitutes normal appearance, and any deviation will be considered a defect. Wilting of leafy vegetables is an obvious defect and therefore unacceptable to the consumer. Skin blemishes, such as bruises, scratch marks and cuts, detract from appearance and in most markets detract from price, even when the blemishes reduce neither keeping quality nor eating quality. Although a premium price may be obtained for produce that is free from blemishes, there will still often be a market for lower grade produce (e.g. through roadside stalls). Consumers in less affluent markets tend to be more accepting of defects through economic necessity. This may result in a greater proportion of a crop that reaches the market being sold and consumed, but a reduced return to the provider or trader.

Acceptable appearance can differ between countries and between different regions within a country. In Japan, great importance is attached to the unblemished appearance of fruit. For example, only 'netted' melons without a 'ground spot' are in demand on Japanese markets.

Thus, appearance is a major determinant of quality, and often it is the only criterion available to the buyer. Taste-testing is rarely practised or encouraged at the retail level, although it is commonplace at the commercial markets from which retailers purchase their commodities.

Mouthfeel

Mouthfeel, including texture, is the overall assessment of the feeling a food gives in the mouth. It is a combination of sensations derived from the lips, tongue, walls of the mouth, teeth and even the ears. Each of these areas is sensitive to small pressure differences and responds to different attributes of the produce. Lips sense the type of surface being presented; for instance, they can distinguish between hairy (pubescent) and smooth (glabrous) surfaces. Teeth are involved in determining rigidity of structure. They are sensitive to the amount of pressure required to cleave the food and to the manner in which the food gives way under the applied force.

The tongue and walls of the mouth are sensitive to the types of particles generated following cleavage by the teeth. The amount of juice released is also assessed. Ears sense the sounds of the food being chewed, intimately complementing mouthfeel. Sound is vitally important in produce such as celery, apple and lettuce, where crispness is a critical attribute. The cumulative effect of these responses creates an overall impression of the mouthfeel of the produce.

Flavour

Flavour is comprised of taste and aroma. Taste is due to sensations felt on the tongue. The four main taste sensations are sweet, salt, acid (sour) and bitter. Each sensation is largely perceived at a specific area of the tongue (Figure 10.1). All food tastes elicit a response in one or more areas. The taste of fruit and vegetables is usually a blend or balance of sweet and sour, often with overtones of bitterness due to tannins. They are not naturally salty. Aroma is due to stimulation of the olfactory senses in the nose by volatile organic compounds, some of which have been outlined in Chapter 2. Pain is now a recognised sensory quality, exemplified by 'hot' chillies which damage the linings on the tongue and in the mouth generally.

Nutritional value

Although it is slowly increasing in importance, nutrition is possibly the least important consideration in determining a consumer purchase, since most nutrients can neither be seen nor tasted. Fruit and vegetables are the sole source of vitamin C in the diet of many people, yet only a minority decide to buy a particular type of fruit or vegetable because it contains more vitamin C than another type. Even so, nutritional properties of crops such as avocado have been successfully promoted as being 'cholesterol free', while oranges are widely perceived as being high in vitamin C.

Figure 10.1 Areas of taste sensation on the tongue

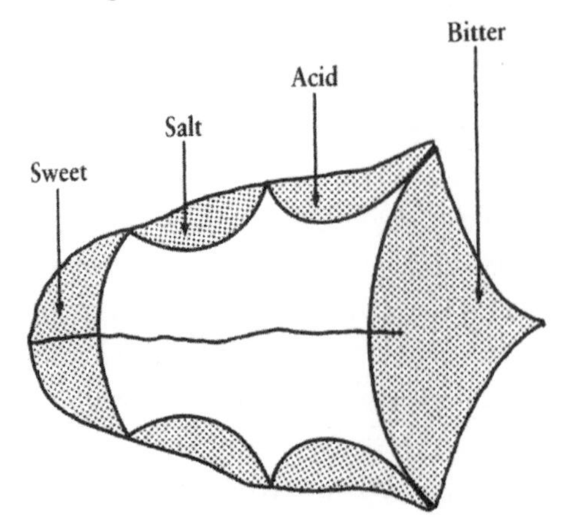

Improved nutritive value is an aim of the fruit and vegetable industry, as it should enhance the health of the community and also lead to increased sales. A joint Australian and New Zealand research and development project called 'Vital Vegetables' is an example of such aspirations. However, when maize varieties with superior nutritional value were produced in the past, acceptance by growers and consumers was marginal. The breeding program developed varieties with high levels of the amino acids lysine and tryptophan, which help overcome a nutritional deficiency associated with normal corn varieties. However, early acceptance and therefore realisation of the health benefit was limited. The new varieties looked different and had some different cultural requirements, so both farmers and consumers saw no immediate advantage in the new corn. However, increased consumer awareness coupled with clever marketing should contribute to more rapid adoption of nutritionally enhanced fruit and vegetable cultivars in the future.

Postharvest factors influencing quality

Not all changes in harvested produce equate to loss of quality. Many of the physicochemical changes occurring after harvest are essential in attaining the desired degree of eating quality. In general, climacteric fruit such as banana and European pear are picked at the mature-green stage of development and then allowed to ripen off the plant to optimum eating quality. Harvest at the mature-green stage is almost mandatory for avocado fruit, which will not ripen while attached to the tree. Avocado fruit does not detach naturally until very late in the season, with virtually concomitant ripening. Nonetheless, the principal concern with many other harvested horticultural products, such as non-climacteric fruit, immature and leafy vegetables and cut flowers and foliage, is preventing loss of the existing quality. Deterioration in quality can be caused by a variety of stresses that may be grouped into four general but often inter-related categories: metabolic stress, transpiration (water) stress, mechanical injury stress, and microbial damage.

Metabolic stress involves either normal or abnormal metabolism, which leads to senescence or the development of physiological disorders, respectively. Respiration-induced carbohydrate shortage in cut flowers and heat treatment-induced impairment of starch breakdown in mango are, in turn, examples of normal and abnormal metabolism. While the loss from physiological disorders is often spectacular, such disorders are generally far less common a problem than normal senescence. Furthermore, certain aspects of general senescence can be just as rapid and dramatic as manifestations of physiological disorders; for example, chlorophyll degradation in harvested broccoli.

Transpiration, the loss of water from plant tissue by evaporation, can also result in rapid loss in quality – and a direct loss in saleable weight. Water loss mainly affects appearance, through wilting and shrivelling, and texture, such as loss of crispness in lettuce. However, water loss can also affect nutritional quality. For instance, vitamin C levels fall rapidly in water-stressed leafy vegetables.

Mechanical injury causes loss of visual quality through unsightly abrasions, bruises, cuts and tears. Such injuries also lead to an increase in the general metabolic rate (wound response) as the produce tries to seal off its damaged tissues. Furthermore, transpiration increases because natural barriers against the loss of water (e.g. the cuticle) have been damaged. Such injury is not always the result of an unintentional action, but may be a side-effect of a deliberate postharvest treatment; for example, damage to the cuticle through the brushing of peaches to remove the fuzz (trichomes) from the skin.

Microorganisms are often considered a secondary stress, since their proliferation is generally facilitated by mechanical injury, transpiration and/ or metabolic changes such as senescence and physiological disorders. This relationship is particularly true for the weak saprophytic organisms that cause postharvest rots, like *Alternaria alternata*. Anthracnose disease of avocado fruit, which is caused by *Colletotrichum gloeosporioides*, becomes evident during ripening when levels of natural antifungal compounds, such as dienes, fall below fungistatic concentrations. Consequently, many microbial problems can be minimised or eliminated by careful postharvest handling practices. Postharvest diseases are mainly caused by fungi, although some bacteria and yeasts are also pathogenic. Where favourable environmental conditions of temperature, pH and water status prevail, the growth of pathogens can be extremely rapid and result in extensive losses (see Chapter 9).

Some of the major handling factors contributing to loss of quality of harvested produce are discussed below.

Harvesting

Mechanical damage during harvesting and associated handling operations can result in defects on the produce and permit invasion by disease-causing microorganisms. The inclusion of dirt from the field can aggravate this situation. Produce can overheat and rapidly deteriorate during temporary field storage. Failure to sort and discard immature, overripe, undersized, misshapen, blemished or otherwise damaged produce creates problems in the subsequent handling and marketing of the produce.

Transport and handling

Rough handling and transport over bumpy roads damages produce by mechanical action. At high temperatures, produce will overheat, especially if there is inadequate shading, ventilation and/or cooling. Transport on open trucks can result in sun-scorch of the exposed produce. Severe water loss, especially from leafy vegetables, can also occur under these conditions. Inappropriate packaging (e.g. overfilling or underfilling) may result in physical damage of produce due to bruising or abrasion as the commodity moves about during transport. Temperature variations can lead to condensation, which may encourage decay and weaken packages.

Storage

Delays in placing produce in cool storage after harvest often lead to a rapid deterioration in quality. Poor control of storage conditions, over-long storage and inappropriate storage conditions for a particular commodity will also result in a poor-quality product. With mixed storage of different commodities, ethylene produced by one product (e.g. ripening fruit) can promote rapid senescence of another product (e.g. leafy vegetables). Storage at temperatures that are too low may induce physiological disorders or chilling injury. High temperature and high humidity can encourage both superficial and internal mould growth and stimulate the activity of infesting insects.

Marketing

A serious reduction in quality can occur in produce displayed for lengthy periods in retail outlets because of poor marketing organisation. Major causes of quality reduction during marketing include ongoing growth (e.g. opening of cut flowers); water loss leading to wilting; undesirable ripening (e.g. softening of apples) and senescence (e.g. yellowing of leafy vegetables) under conditions of poor temperature and RH management; mechanical damage associated with rough handling by staff and customers; and associated disease development. An interesting marketing problem is the greening of potatoes, which is caused by the practice of displaying cleaned potatoes in relatively bright light in supermarkets. Greening is strongly associated by consumers with accumulation of the toxic glycoalkaloid, solanine. The desirable introduction of refrigerated sections in retail outlets can, however, bring their own problems if, for example, chilling-sensitive produce are held at low temperatures.

Treatment residues

Residues of pesticides and other chemicals are an increasingly important factor affecting postharvest quality. Chemicals such as insecticides and herbicides are often applied preharvest. Fungicides may be used both preharvest and/ or postharvest to prevent rotting (Chapter 9). Fumigants may be used for insect disinfestation, especially when exporting, or for disease control. All applied chemicals can leave residues in the commodity. Although these are usually not detectable by the consumer, the health risks to the community must be considered. The use of electron beam and gamma irradiation to inhibit mould growth or sprouting can often reduce the need for chemicals, but these methods have their own image problems with consumers.

Determining maturity

Maturity is an integral component of quality, especially in the context of commercial maturity. There is a clear distinction between 'physiological' and 'commercial or horticultural' maturity. The former is a particular stage in the development of a plant or plant organ, and the latter is concerned with the time of harvest as related to a particular end-use that can be translated into market requirements. At optimum commercial maturity, produce should be either at optimum consumer quality (e.g. ripe in the case of non-climacteric fruit such as strawberry) or able to achieve optimum consumer quality (e.g. at an advanced bud stage in the case of potted chrysanthemum plants intended for the Mother's Day market).

Physiological maturity refers to the point in the development of a plant or plant part when maximum growth has been achieved and it has matured to the extent that the next development stage can be completed. In the case of fruit, ripening can be considered the next development stage, preceding the senescence stage. It is generally not easy to clearly distinguish between the stages of development – growth, maturation, ripening and senescence – in a plant organ or organism (see Chapter 3). This is because transitions between the various development stages are often slow and/or indistinct. Nevertheless, in fruits in particular, physiological (e.g. respiration and ethylene production) and/or biochemical characteristics (e.g. sugar–acid ratios) can be measured to give reliable estimates of the degree of maturity for specific commodities (see Chapter 11).

Commercial maturity is the characteristic state of a plant organ required by a market. This often bears little relation to physiological maturity, and may occur at any stage during development, maturation, ripening or senescence. Examples of commercial maturity include bean sprouts (during development), cucumber (during maturation) and strawberry (during ripening/senescence). The terms immaturity, optimum maturity and over-maturity can be related to these market requirements, so there must be understanding of each in physiological terms, particularly where storage life and quality when ripe are concerned. Some examples of commercial maturity in relation to physiological age are shown in Figure 10.2.

Figure 10.2 Commercial maturity in relation to developmental stages of the plant

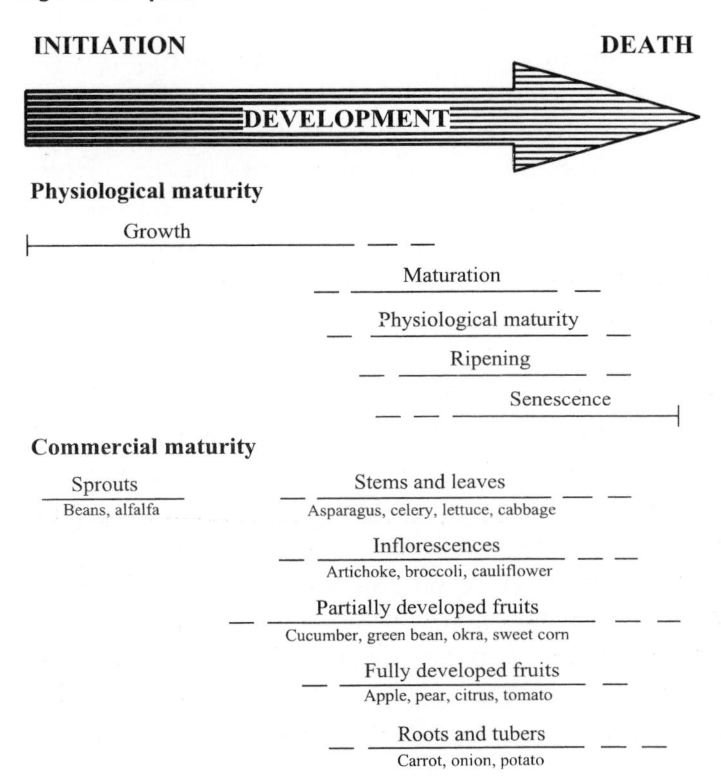

SOURCE Adapted from A.E. Watada, R.C. Herner, A.A. Kader, R.J. Romani and G.L. Staby (1984) Terminology for the description of developmental stages of commercial crops. *HortScience* 19, pp. 20–21.

Ripening, as applied to fruit, is the process by which it attains its maximum desired eating quality. The ripening process presumably evolved as a means of attracting fruit-eating animals in order to ensure dispersal of the seeds within fruit. Before the ripe stage, a fruit is said to be under-ripe; afterwards, it is over-ripe. The three ripeness conditions cannot be clearly defined physiologically because they are subjective judgements and will vary among consumers. The under-ripe condition overlaps the mature stage of development, while the over-ripe condition overlaps the senescence stage of development. Consider the pear, which some relish at a very soft succulent stage and others prefer at a crunchier stage. The potential ripe quality is determined by many factors, but the stage at which the fruit was harvested is vitally important.

Determining commercial maturity

Commercial maturity indices generally involve some expression of the stage of development (growth, maturation or ripening) and usually require determination of some characteristic known to change as the plant material develops. They may involve taking decisions about levels of market and consumer acceptability, and generally necessitate making objective measurements, subjective judgements or both. Objective and subjective assessments may be destructive or non-destructive in nature.

Determining the time to harvest peas for processing by canning or freezing provides an interesting example of the need for close monitoring of commercial maturity. Destructive objective measurements of maturity in shelled peas can be obtained using several specially designed instruments that measure the force required to shear the seed. The end result is provision to the consumer of a uniform, high-quality canned or frozen product.

Such objective standards cannot always be used to determine maturity in commodities. The time of harvest is often judged by growers based on their experience with their own crops in terms of calendar date, and various subjective judgements in relation to market requirements. Cut roses are a good example. Those destined for local markets can be harvested at a more advanced stage, with the outermost petal lifting from the bud, while those destined for storage and transport to distant markets are harvested at a tight bud stage.

Many criteria for judging maturity are based on a variety of characteristics, including: time from flowering or planting (calendar date); accumulated heat units; size and shape; skin or flesh colour; light transmittance or reflectance; flesh firmness; electrical conductance or resistance; chemical composition (e.g. starch, sugar, acid); respiratory behaviour and ethylene production; and time to ripen (Table 10.1). The ideal maturity tests are simple, rapid, and readily applied in the field. They are ideally non-destructive. A description of some methods follows.

Table 10.1 Established methods for evaluating maturity in horticultural produce

Maturity index	Produce
Abscission	Rockmelon
Accumulated heat units	Pea
Astringency	Persimmon
Chronological time	Apple
Colour	Tomato
Stage of development, e.g. density	Lettuce
Dry matter content	Mango
Firmness	Pea
Internal ethylene	Apple
Juice content	Orange
Oil content	Avocado
Shape	Banana
Size	Gherkin
Starch-iodine staining	Apple
Soluble solids/acid ratio	Orange
Waxiness and gloss	Grape

Calendar date

For perennial fruit crops grown in distinctly seasonal climates that are relatively uniform from year to year, calendar date for harvest can be a useful guide to commercial maturity. This approach relies on a reproducible date for the time of flowering and a relatively constant growth period from flowering through to maturity. Time of flowering is largely dependent on temperature, and the variation in number of days from flowering to harvest can be calculated for some commodities by using the degree-day concept (see below). Calendar-based harvest dates are usually derived from growers' experience with crops in a specific environment.

Heat units

The time required for the development of fruit to maturity after flowering can be calculated by measuring the cumulative 'degree days' or 'heat units' in a particular environment. It has been found that a characteristic number of heat units or degree days is required to mature a crop. Under unusually warm conditions, maturity will be advanced and under cooler conditions it will be delayed. The number of degree days required for a crop to reach maturity is determined over a period of several years from the sum of the differences between the daily mean temperatures and a fixed base temperature (commonly the minimum temperature at which growth occurs) over the whole growing period. The total number of degree days over the growing season (e.g. 1000 deg.d) is then used to forecast the probable date of maturity for the current year. This heat unit approach is extremely helpful in planning planting, harvesting and factory management programs for crops that are processed annually, such as corn, peas and tomato. This approach is also useful in programming cut flower or flowering pot plant production (e.g. lilies for the Easter market, poinsettias for the Christmas market).

Shape and size

In some instances fruit shape may be used to decide maturity. For example, the fullness of the 'cheeks' adjacent to the pedicel is a guide to maturity of mango and some stone fruits. Size is generally of limited value as a maturity index in fruit, though it is widely used for many vegetables – especially those marketed early in their development, such as zucchini. Size, in combination with shape, is also an important determinant of commercial maturity in ornamentals, e.g. flowering and foliage pot plants. With such produce, size is often specified as a quality standard. Large size generally indicates commercial over-maturity and under-sized produce indicates an immature condition.

Shape and size give rise to the volume parameter, which when expressed as a function of weight (mass) gives density (e.g. g/mL). This ratio is a useful determinant of maturity in some produce (e.g. potato) and to grade frost-damaged citrus fruit. It can also be used to sort defects (e.g. disorders that result in internal cavities). Current commercial sorters use load cells to weigh fruit, and cameras with multiple views of fruit (obtained by using mirrors or rolling fruit) to estimate volume – and thus can calculate density.

Colour

The loss of green colour, often referred to as the 'ground colour' (i.e. the background colour), is a valuable guide to maturity in many fruits. Initially there is a gradual loss of colour intensity from deep green to lighter green, and many commodities completely lose the green colour and develop yellow, red or purple pigments (e.g. tomatoes). Ground colour, as judged against prepared colour charts, is a useful index of maturity for apple, pear, stone fruits and mango, but is not entirely reliable because it is influenced by factors other than maturity (e.g. nitrogen nutrition during growth).

For some fruits, additional colour superimposed on the ground colour can be a useful indicator of maturity. Examples are the red or red-streaked apple cultivars and red blush on some cultivars of peach. Such colour development usually depends on exposure to sunlight. For certain tropical fruits, skin colour is a reliable guide to commercial maturity. For example, the appearance of a trace of yellow at the apical (distal) end of papaya fruit is sometimes used to determine the time of harvest. Nonetheless, such fruit would benefit from a longer period on the tree, until about one-third of the fruit is yellow. After this time, papaya fruit ripens quicker and eating quality is better, but shelf-life is shorter. Thus, the grower must balance the benefits of improved eating quality against greater risk of losses associated with over-ripening during marketing. For table market tomatoes, both the loss of green colour (chlorophyll) and the development of red colour (lycopene) are indicated in pictorial charts by which maturity is judged (Plate 6).

Colour can be measured objectively using a variety of light reflectance or transmittance spectrophotometers. The Hunter and Minolta colour/colour difference meters are widely used in research work. Colour video image analysers can also be used to sort horticultural produce on the basis of both colour and dimensions (Plate 4). Electronic colour vision sorting is now common for many crops (e.g. apple, citrus and tomato). As the capital cost of sophisticated equipment is high, its use is largely restricted to packing houses with high throughput of produce. Nonetheless, as labour costs increase, broader (e.g. field) application of such equipment can be anticipated.

Changes in internal colour and other physicochemical characteristics can be gauged by measuring the transmittance/absorbance of light through a sample. Such measurements can be made rapidly, at the speed of modern fruit conveyors. For example, internal green colour, a defect of gold kiwifruit, can be readily assessed. Grading of fruit for internal chemical properties, such as sugar level, is possible using near infra-red spectroscopy (Figure 10.3).

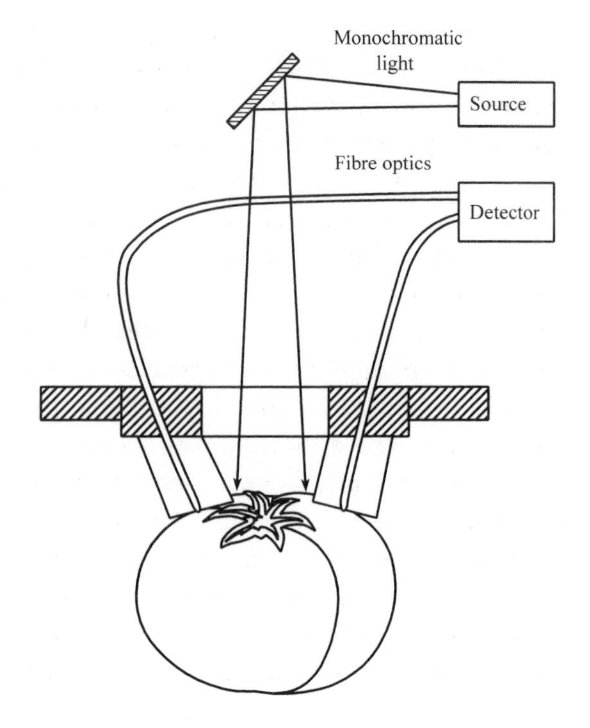

Figure 10.3 Diagram of the geometry used to make transmission measurements of fruit. The incident light that enters the produce directly under the source is scattered and absorbed by the produce. Some of the scattered light transmitted by the produce is collected by the shielded detectors placed directly on the surface of the produce. Transmission spectra are generated by varying the wavelength of the incident light.

SOURCE Adapted from G.S. Birth, G.G. Dull, J.B. Magee, H.T. Chan and G.G. Cavaletto (1984) An optical method for estimating papaya maturity. *Journal of the American Society for Horticultural Science* 109, pp. 62–69.

Flesh firmness

As fruits mature and ripen, they soften, largely by dissolution of pectins comprising the middle lamella of cell walls. This softening can be estimated subjectively by finger or thumb pressure. However, more objective measurement, yielding a numerical expression of flesh firmness, is possible with a fruit pressure tester (penetrometer). These testers measure the pressure at which flesh yields to the penetration of a standard diameter plunger inserted to a standard depth. Commonly used penetrometers are the Magness-Taylor and UC Fruit Firmness testers, and the smaller and more convenient Effegi penetrometer (Figure 10.4). While the comparatively inexpensive penetrometers do not necessarily give the same numerical value if used on the same produce, each instrument will give reproducible values for the purpose of comparative evaluation. Therefore, it is necessary to specify the instrument used when reporting pressure test values or attempting to set standards.

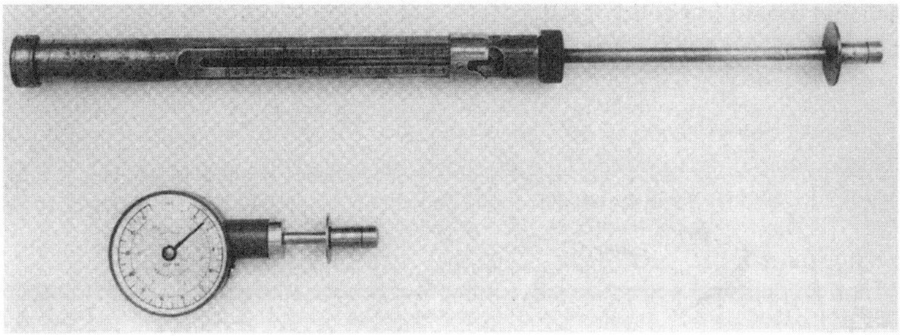

Figure 10.4 Magness-Taylor (top) and Effegi (bottom) fruit pressure testers

Chemical measurements

Measuring the chemical characteristics of produce is an obvious approach to the problem of determining maturity, particularly as they can often be directly related to taste (e.g. sweetness, sourness) of produce. The conversion of starch to sugar during maturation is a simple test for the maturity of some apple varieties. It is based on the reaction between starch and iodine to produce a blue or purple colour. The intensity of colour indicates the amount of starch remaining in the fruit. This test can also demonstrate the disappearance of starch from the pulp of ripening banana and the petals of ageing rose flowers. Sugar is usually a major component of soluble solids in cell sap and can be measured directly by chemical means. However, it is much easier and just as useful to measure the concentrations of soluble solids in extracted juice with a refractometer (Plate 7) or hydrometer. These devices measure the refraction of light as it passes through a thin sample of juice containing sugars, and variation in the density of juice according to its sugar content, respectively. Maturity standards for melons, grape and citrus are often based on soluble solids concentration.

Acidity (titratable acidity) can be determined for a sample of extracted juice by titrating it with an alkaline solution (normally 0.1 N NaOH) to a colour change of a pH indicator (e.g. phenolphthalein) or to a specific pH (commonly 8.1). Loss of acidity during maturation and ripening is often rapid. The sugar/acid or total soluble solids/acid ratio is often better related to palatability of the fruit than to either sugar or acid levels alone. Maturity standards for citrus fruits are commonly expressed as a ratio of the % soluble solids to level of acid, both measurements being on a weight for weight basis. In Figure 10.5 the increase in the soluble solids to acid ratio is shown for Valencia and Washington Navel orange fruit maturing on the tree. A minimum soluble solids concentration is often included in specifications of citrus and pineapple fruits for processing and of grapes used for drying or juice production.

Figure 10.5 Variation in soluble solids/acid ratio in juice from Washington Navel oranges grown on trifoliata rootstocks

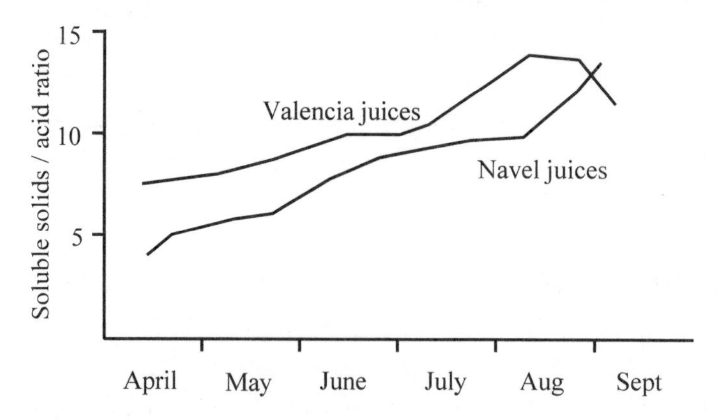

SOURCE Adapted from B.V. Chandler, K.J. Nicol and C. von Biederman (1976) Factors controlling the accumulation of limonin and soluble constituents within orange fruits. *Journal of the Science of Food and Agriculture* 27, pp. 866–76.

Dry matter content, which tends to increase as certain fruits mature, can be used for produce in which there is a large increase in the amount of starch or sugar during maturation (e.g. kiwifruit and mango). Although dry matter content is most widely used for avocado, oil content can also be used as a maturity index for this compositionally unusual fruit.

Climacteric behaviour

Commercial maturity can be related to the rise in respiration rate and/or ethylene production in climacteric fruits. For longest storage and optimum eating quality, apples and pears should generally be picked just before the climacteric rise. If they are picked too early in the preclimacteric period, their ripened quality will be poor. In practice, the appropriate point on the respiration curve must usually be related to some other maturity characteristic that can be readily determined in the field (see above). In the case of once-over (strip) picking, harvesting might be commenced when a certain small proportion of the fruit have commenced ripening.

The term 'green-life' has been coined to describe the degree of physiological immaturity at the time of harvest in terms of the time before the fruit begins to ripen. Less-mature fruit has a longer green life than more mature fruit, e.g. for avocados, bananas and mangoes. This concept is useful for expressing potential postharvest life.

Novel technologies for determining maturity and/or defects

Meaningfully determining maturity is inherently difficult, not least because maturity is part of a plant or organ development continuum. That is, maturity has no discrete beginning or end. Thus, researchers are continually looking for better or improved ways to determine maturity. Similarly, the search for cost-effective methods of detecting defects, particularly internal defects, is continuing. Some novel non-destructive technologies that may in time find practical application include chlorophyll fluorescence spectrophotometers that can detect chlorophyll loss and X-ray and magnetic resonance systems that are based on detecting differences in tissue density or proton mobility in tissue, respectively. The latter techniques can, for example, reveal the commencement of ripening in the inner mesocarp and/or internal cavities associated with the activity of fruit fly larvae in mango fruit (Plate 8). Like common measures of visible colour, X-rays and magnetic resonance (radio frequencies) utilise parts of the electromagnetic spectrum (Figure 10.6). Near infra-red spectroscopy utilises yet another portion of this spectrum, and is finding increasing application as a means for non-destructively determining soluble solids concentration in relation to harvest maturity or eating quality of crops such as apples and stone fruit. Molecular probes that bind to mRNA or proteins might be used to detect early biochemical changes during ripening (e.g. synthesis of ACC synthase or ACC oxidase). Finally, solid state broadband (i.e. chemical class, such as aldehydes) sensors or biosensors may be applied to 'sniff' produce in order to detect the presence of volatiles associated with ripening or disorders (e.g. decay).

Figure 10.6 Electromagnetic spectrum illustrating the wavelengths of radiation that can be used for the non-destructive detection of changes in the structure and composition of plant tissues

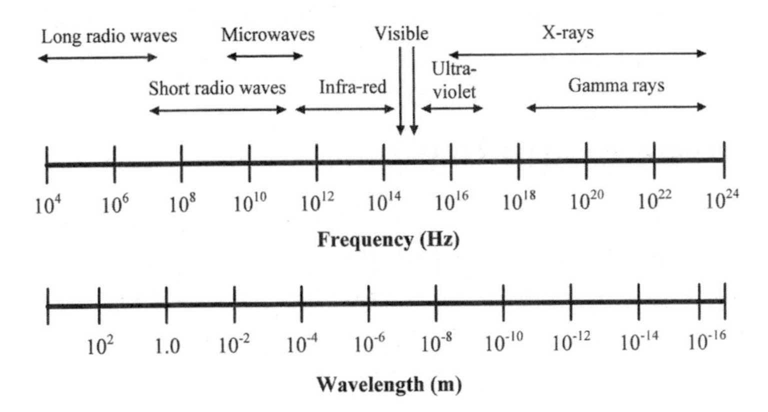

Quality evaluation for ornamentals

Postharvest quality testing of ornamentals merits particular attention. Many people readily associate 'postharvest' with fruit and vegetables, but not so readily with postproduction handling of ornamentals – especially of whole plants. This misconception also applies to vegetable and ornamental seedlings, propagules such as bulbs and corms, cuttings and tissue cultures, and advanced nursery stock. Nonetheless, there is increasing awareness of the need to define and test postproduction quality of ornamentals. In Europe, in particular, the ornamentals industry undertakes postproduction quality evaluation, largely because the industry is extremely competitive and consumers are highly discerning. The Netherlands has traditionally led the way in applied quality testing. Inspectors sponsored by the major flower auctions in Holland assess the quality of every consignment received. These same auctions also operate laboratories that conduct vase life and quality loss assessments on cut flowers and pot plants, respectively. Typical vase life assessment room conditions are 20°C and 50–70% RH, under an artificial fluorescent light regime of 10–20 $\mu mol/m^2.s$ at flower level for 12 hours per day. These conditions simulate the indoor home environment. Typical test conditions for pot plants are simulated transport (4 days at 16°C and 90–95% RH) followed by shelf life evaluation (14 days at 20°C, 55–60% RH and 8 $\mu mol/m^2.s$ for 12 hours per day). Largely subjective evaluations (i.e. based on visual appeal) are undertaken by trained professional staff employed by the auctions. Increasingly, growers and private consultants on behalf of growers or other interested parties also conduct quality evaluation. Many quality loss assessments only provide feedback after the particular consignment has been sold to consumers. With this limitation in mind, chlorophyll fluorescence parameters have been related to postproduction performance in an effort to develop a rapid and reliable predictive assessment of quality.

Management of quality

The gradual breakdown of barriers to free international trade has led to the concept of the 'global village', in which fresh produce is able to move freely from country to country based on seasonality of supply and climatic conditions. Because of this, domestic markets in many countries are facing increased competition from imported products. However, global trade is not new. For many years, tropical fruits such as banana produced in areas like the Caribbean were transported by sea to Europe and North America. Similarly, large volumes of cut flowers grown in countries such as Colombia are air-freighted to Europe and North America, and apples grown in temperate areas of the world are exported to tropical countries. The freeing of international trade has led to a much stronger emphasis on quality assurance so that buyers can rely on produce specifications being consistent.

Each market has its own criteria for home consumption and for export, depending on local circumstances, but generally only the higher quality lines are exported, because of the longer time exported produce has to last before consumption. Aspects of quality such as those outlined in the previous section are today considered primarily as commercial issues, set by customers in order to procure the standard of produce required for an individual market. In the past, specifications were often statutory requirements administered and enforced by government agencies, but today the import customers, often large retailers, drive the development of produce specifications in most countries, since they are usually the final point of contact with consumers. These specifications clearly define a set of produce attributes, packaging and labelling requirements and, increasingly, state the acceptable quality/food safety assurance systems under which produce must be grown and handled. In specifying system requirements, the customer establishes the mechanism through which their specifications may be managed and enforced throughout the supply chain. While the focus continues to be on visual assessment, and advanced optical systems are now widely used for sorting produce for size, colour and surface blemishes, advances in technology for the non-destructive testing of internal quality attributes now allow far more accurate prediction of quality aspects such as flavour and texture.

Government activity is now primarily restricted to the assurance of pest- and disease-free status for both domestic and export markets. For many years there has been a requirement for fresh produce moving between countries to be guaranteed free of certain pests and diseases. The governments of exporting countries are responsible for providing an inspection service and issuing the necessary phytosanitary certificates required by importing countries, as well ensuring that imported produce does not breach domestic quarantine requirements. Within Australia, interstate certification is also required for some produce moving into areas with pest-free status.

Quality systems

While quality may be considered a commercial factor, other aspects of produce assurance such as food safety are today considered fundamental to supply. The primary non-commercial element of product assurance is food safety, with consumers expecting a guarantee that fresh produce does not contain pesticides or microorganism residues that could harm human health. In the coming years, issues such as occupational health and safety (OH&S), worker welfare, biosecurity and bioterrorism will all become increasingly important 'tickets' to supply, providing for a 'total on-farm assurance package'.

Since the early 1990s the horticultural industries have moved towards a holistic system in which quality is managed along the whole distribution chain from the farm or orchard to the final point of sale. To achieve this, it is necessary to monitor and prevent quality problems as early as possible in the production process, rather than relying on endpoint and reactive inspection at a later stage during distribution when the product has become more valuable. Quality assurance is especially important when produce is being shipped long distances over long periods to overseas markets. Though implementing quality assurance systems may entail some cost, a well-managed system will ultimately reduce costs by preventing quality problems, and the supplier will gain a marketing advantage by providing consumers with confidence that the product will consistently meet their specifications.

The early quality assurance systems in horticulture evolved from formal quality management systems/structures in other industries. The ISO 9000 series was used as the basis for early attempts at system development, but the rate of system uptake was slow. Today, the focus is on a risk-management-based approach, with the majority of the systems underpinned by the HACCP (hazard analysis and critical control points) technique. While conventionally used for food safety, HACCP is increasingly being used across a range of 'assurance areas', enabling individuals to assess risk and thus identify what might go wrong, establish controls to minimise the likelihood of such an occurrence and take corrective actions to manage those that do occur. The formal steps of HACCP are as follows:

1. Identify and assess all hazards
2. Identify the critical control points
3. Identify the critical limits
4. Establish the monitoring procedures

5. Establish corrective actions
6. Establish a record-keeping system
7. Establish verification procedures.

An example of the application of HACCP principles to pesticide residues is given in Table 10.2. This approach has removed much of the 'fear' and 'confusion' traditionally associated with quality assurance systems. The prescriptive format clearly documents what needs to be done and when and what records need to be kept.

Table 10.2 Example of the application of HACCP principles to prevent a potential contamination of harvest produce with chemical residues in excess of the approved maximum residue level (MRL)

Hazard	Critical control point	Cause of hazard	Preventative measure	Critical limit	Monitoring	Corrective action
Pesticide residue application	Yes	Insufficient harvest interval	Adhere to withholding period prior to harvest	Withholding period stated	Application and harvest dates	Spray records
Exceeding MRL application	Yes	Incorrect application rate of chemical	Apply chemical at recommended rate	Recommended rate	Spray records	Spray records

Examples of this approach include EurepGAP, which as a HACCP-based prescriptive code has in recent years become the recognised standard for growers supplying product into the European market; or the BRC (British Retail Consortium) Standard, a prescriptive code developed for the supply chain post–farm gate. In the Australian domestic market, a number of such programs exist within the fresh produce industry, including Freshcare – a program developed by the Australian horticultural industry to facilitate the management of on-farm food safety and latterly, to encourage broader on-farm assurance. An industry-owned and run program, Freshcare is a prescriptive Code of Practice, based on a master HACCP plan. Another HACCP-based system is the SQF 2000™ and SQF 1000™ (Safe Quality Food) codes, originally developed by Agriculture Western Australia, a department of the Western Australian Government, and now owned by FMI (Food Marketing Institute) in the USA.

A key factor in the successful uptake of any assurance program is the mode of delivery. Experience has shown that greater grower involvement in training and in system implementation leads to a more successful outcome. In general, where systems are developed and implemented by an external consultant without focussed client training and involvement, the ability of the client to successfully maintain that system is limited.

Most horticultural producers today find that the record-keeping required to maintain their chosen system is second nature and simply part of good business practice. Another important element in achieving and maintaining customer confidence is auditor consistency. Most customers require third-party (independent) certification to a nominated system or standard. The majority of the 'System/Standard Owners' have clear requirements for auditor capability and experience.

Future developments

With food safety and quality systems now a fundamental requirement for supply to most markets, the focus is on broadening the scope of on-farm assurance systems to encompass areas such as environmental assurance, biosecurity, and occupational health and safety.

The increasing number of assurance programs worldwide is also driving the need for a recognition framework to be developed, whereby individual standards organisations can be benchmarked to enable inter-country recognition without businesses requiring multiple certifications. This process has been initiated in part by programs such as EurepGAP and the Global Food Safety Initiative (GFSI) but there is still a long way to go before the global fresh produce industry can claim to have established consistent requirements and implementation across all markets.

Supply chains in an increasingly global horticultural marketplace will continue to strive for assurance systems that will enable participants to maintain existing market share, achieve access to new markets and deliver consistent performance from their businesses in meeting customer expectations.

11 Preparation for market

Preparing produce for market covers handling operations from harvest until produce leaves the packing house. The main operational areas are harvesting, postharvest treatments, grading and packaging. Control of diseases (Chapters 8 and 9), and evaluation and management of quality through to the retail level have been discussed earlier (Chapter 10), while storage recommendations are discussed later (Chapter 13).

Harvesting

While the harvest process for potted plants may simply require transferring plants from their production environment, harvesting fresh fruit and vegetables and cut flowers generally involves separating them from their vital sources of water, nutrients and growth regulators. Harvesting may also elicit wound responses (e.g. ethylene production, increased respiration) in the tissue. Mature fruit generally shows only a small response to harvesting because it has stored carbohydrate reserves and relatively low respiration and transpiration rates, and is destined for natural separation by abscission anyway. Rapidly metabolising tissues (e.g. leafy vegetables) exhibit larger responses to harvesting.

It is usually best to conduct harvest operations in the morning when the heat load is low. Around-the-clock mechanical harvesting is carried out for some processing crops (e.g. beans) in order to make best use of expensive machinery and to meet factory processing schedules. Night-harvesting may also be conducted for field-grown crops, such as fresh market melons.

With some crops it is prudent to wait until the early morning dew is gone and some fruit turgor is lost before harvesting. This is particularly so for oranges as, when fully turgid, the oil cells of the flavedo (skin) readily rupture under applied pressure (e.g. other fruit in a bin) and spill their toxic contents, causing skin discolouration and injury. Similarly, certain mango varieties, such as Kensington Pride, are more prone to sap damage when fully turgid. Mango sap stains and is also caustic. In some rapidly respiring crops with little or no carbohydrate reserves (e.g. cut flowers) there may be a case for harvesting late in the day when carbohydrate levels are highest. However, in the absence of significant storage capacity, the advantages of an early morning harvest predominate.

Harvest mechanisms include breaking off (e.g. twisting-off pineapples by hand), cutting (e.g. snipping-off mandarins and table grapes with secateurs) and shaking (e.g. vibrating-off almonds with mechanical tree-trunk shakers). The mechanism chosen largely depends on the strength of attachment of the organs and the degree of damage that is tolerable, in addition to economic considerations. For fruit such as olives, weakening the attachment of fruit by applying a chemical loosening agent (ethephon, which breaks down to liberate ethylene) facilitates harvesting. For fruit on high trees, such as avocados, picking poles fitted with a cutter device that can grip the fruit and sever the stalk may be used. Motorised 'cherry pickers' with extendable arms that can support an operator and a picking bag are an extremely useful, albeit expensive, picking aid.

The harvest operation may be wholly by hand or by machine. Hand harvest predominates for fresh market produce, particularly produce that is susceptible to physical injury (e.g. apples, roses). In regions where labour costs are high, machine harvest is popular for processing crops (e.g. peas for freezing, peaches for canning, grapes for winemaking). Similarly, mechanical harvesting is used for robust, low-unit-value ground crops such as potatoes and onions. Combination processes make use of mechanical aids. For example, conveyors are used to move hand-harvested pineapples and melons from the rows of plants to bins mounted on a trailer moving along parallel vehicle access aisles. Novel harvest aids include platforms on which people lie or sit to gather strawberries, and 'flying foxes' (overhead ropeways) to convey heavy banana bunches into the packing house. Machines are also used in harvesting rows of trellised tomatoes; pickers place hand-harvested fruit onto conveyers that elevate fruit into bins carried on the machine.

In some production systems, harvesting, packing and even cooling are all undertaken in the field. For example, lettuce grown on very large farms in California may be cut and elevated to a packing platform for cleaning, trimming and packing. The packed cartons may then be cooled in mobile vacuum coolers.

An important precaution at harvest is to avoid contaminating produce with pathogens. Practices such as standing mangoes stem-end down on the ground to allow the sap to drain should be discouraged. Another important precaution is to avoid physical injury. Ways to avoid physical injury include deceleration chutes to transfer fruit from picking bags to bins, lining picking bins and farm trailers with padding, and maintaining smooth farm roads. Harvested produce should be kept shaded either by natural (e.g. tree canopy) or artificial means (e.g. tarpaulins).

Postharvest treatments

Many postharvest treatments are applied to horticultural crops, either to maintain quality or to improve visual appeal. Harvested produce must be handled with care at every stage to avoid mechanical damage. For instance, large drop heights, friction areas and sharp objects should be avoided.

Washing

Washing may be important to remove sap (e.g. mango), soil (e.g. carrot) and debris (e.g. banana). Clean water is essential, otherwise fungi and bacteria may build up. A series of two or three washes may be beneficial, possibly with an approved disinfectant treatment (e.g. chlorine, iodine) applied to the last wash (Plate 9). Soap and other chemicals, such as calcium hydroxide, can also be added to washing water to facilitate the process.

Waxing

Commodities with a waxy skin tend to lose water slowly. This observation has led to the application of wax to certain commodities that shrivel rapidly and lose consumer appeal during storage and marketing. In addition to reducing water loss, waxes are also applied to enhance the appearance of produce to the consumer (e.g. shiny apples). The rate of water loss can be reduced by 30–50 per cent under commercial conditions, particularly if the stem scar and other injuries are coated with wax. Citrus fruit is commonly waxed, because washing can remove much of the natural wax from the peel, thereby exacerbating shrivelling and loss of appearance. Many other commodities, such as cucumber, tomato, passionfruit, peppers, banana, apple and some root crops, are also waxed to reduce weight loss and to increase sales appeal.

Most waxes in commercial use are proprietary formulations. They may contain a mixture of different waxes derived from plant and/or petroleum sources. Many have been based on a combination of paraffin wax, which gives good control of water loss but a poor lustre to the produce, and carnauba wax, which imparts an attractive lustre to the produce but provides poorer control of water loss. In addition, formulations containing polyethylene, synthetic resin materials, sugars and sugar derivatives, chitosans and emulsifying and wetting agents have been used. The wax formulation may be used to carry fungicides and inhibitors of senescence, superficial scald and sprouting. Waxes are brushed, sprayed, fogged or foamed onto produce, or produce is conveyed through a tank of wax emulsion. The wax film must be thin so that gas exchange is not overly hindered, which can cause anaerobiosis and associated quality loss such as off-flavours. After wax is applied, produce is generally dried and polished.

Curing

Underground storage organs such as potato and sweet potato tubers tend to have poorly developed cuticles. Thus they are relatively susceptible to mechanical wounding during harvesting and handling, and to postharvest water loss and decay. These problems can be minimised by the process of 'curing' at intermediate to high temperatures and at high humidity. For example, sweet potatoes can be cured at around 30°C and 90–95% RH for 4–7 days prior to storage. Similarly, potatoes may be cured at 10–15°C and about 95% RH for a period of 10–14 days. During the curing period, a surface layer of protective suberised wound periderm tissue (Chapter 2) is formed over the product, especially at wound sites. Although periderm forms on potato most rapidly at about 21°C, the risk of decay at this comparatively high temperature is unacceptable. Curing treatments may also facilitate wound healing for certain fruits, such as citrus.

Plant growth regulators

The five major types of plant growth regulators are auxins, gibberellins (GAs), cytokinins, abscisic acid (ABA) and ethylene. These phytohormones regulate many plant functions and biosynthetic pathways and have a role in most aspects of plant development, from germination and bud formation to senescence. Postharvest information on the roles of the first four types of growth regulators is not as extensive as for ethylene. However, they all merit ongoing study because of their coordinating roles (at extremely low concentrations) in regulating physiological processes.

Despite their potency and natural occurrence, few postharvest applications of plant growth regulators, except ethylene (see below), have been implemented by industry. Two common commercial examples are retarding button senescence in citrus fruit using 2,4-dichlorophenoxyacetic acid (2,4-D) in order to prevent stem-end rots, and retarding senescence of citrus peel using gibberellic acid (GA_3). Laboratory studies have shown that cytokinin application can delay the loss of chlorophyll from, and the general senescence of, many green leafy materials (e.g. broccoli florets, alstroemeria leaves). Externally applied gibberellins can reduce ethylene production in some cut flowers. ABA can be applied to help maintain the water balance, through induced stomatal closure, of cut flowers and pot plants.

Sprout inhibitors

The buds of potato and onion enter a dormant state at maturity. These vegetables are normally harvested at this time and can be stored for many months under correct conditions for either retail marketing or processing. The duration of postharvest dormancy (rest period) of potato tubers is influenced by preharvest factors, maturity and variety, but generally not by the temperature of storage. Once the rest period ends, the rate of sprouting depends on temperature. Potato rarely sprouts below 4°C, but storage at these temperatures is impractical due to the conversion of starch to sugars (Chapter 4). Sprouting at temperatures greater than 4°C is a problem during storage periods longer than 2–3 months.

The application of several chemicals and ionising radiation (see later in this chapter) can effectively suppress sprouting in potato and onion during storage at higher temperatures. However, legal restrictions on the usage and permitted residues of these compounds vary among countries. 3-Chloroisopropyl-N-phenylcarbamate (CIPC) is a strong sprout inhibitor and probably the most widely used on potato. CIPC may be applied as a dust, water dip, vapour or aerosol. Like many of the other chemical sprout inhibitors, CIPC interferes with periderm formation, and thus should be applied after curing (see above). Sprouting of onions during long-term storage is effectively prevented by applying maleic hydrazide several weeks before harvest.

Disinfestation

Some insect species, such as the tephritid fruit flies that infest a broad range of horticultural fruit crops, pose a biosecurity threat and can seriously disrupt trade in produce between countries. The presence of these insects may also restrict movement of produce within a country. Table 11.1 lists the more important pest species, their distribution and hosts (see also Plate 10). Importing countries often place a quarantine barrier on produce from an area in which the insect species of concern is known to occur. In order to market produce, exporting nations with quarantined insect pests must develop an effective disinfestation treatment that satisfies the importing country by virtually complete destruction of the egg, larva, pupa and adult development stages. However, the disinfestation treatment must not harm either the produce or the consumer, and should be economical to apply.

Table 11.1 Some insects and mites that can be carried by fruit and vegetables

	Common name	Common hosts*	Approximate distribution
Fruit flies			
Anastrepha fraterculus (Wied.)	South American fruit fly	Peach, guava, citrus, *Spondias* spp., *Eugenia* spp.	South and Central America, West Indies
A. ludens (Lw.)	Mexican fruit fly	Citrus, other tropical and subtropical fruits	Central America, Mexico
Ceratitis capitata (Wied.)	Mediterranean fruit fly	Deciduous and subtropical fruits, especially peach and citrus	Southern Europe, Africa, Central and South America, Australia, Hawaii
C. rosa (Karsch)	Natal fruit fly	Many deciduous and subtropical fruits	Africa
Dacus ciliatus (Lw.)	Lesser pumpkin fly	Cucurbits	Africa, India, Pakistan, Bangladesh
D. cucurbitae (Coq.)	Melon fly	Cucurbits, tomato	Asia, Hawaii, Papua New Guinea, Africa
D. dorsalis (Hend.)	Oriental fruit fly	Most fleshy fruits or vegetables	Asia, Hawaii
D. tryoni (Frogg.)	Queensland fruit fly	Many deciduous and subtropical fruits	Australia, Pacific Islands
Rhagoletis cerasi (L.)	Cherry fruit fly	Cherry, *Lonicera* spp.	Europe
R. cingulata (Lw.)	Cherry fruit fly	Wild and cultivated cherry, *Prunus* spp.	North America
Mites			
Halotydeus destructor (Tucker)	Red-legged earth mite	Leafy vegetables	Australia, New Zealand, Africa
Panonychus ulmi (Koch)	European red mite	Apple and other deciduous fruits	Europe, Africa, Asia, Australia, New Zealand, North and South America
Phthorimaea operculella (Zell.)	Potato tuber mite	Potato, tomato, eggplant	Worldwide

Mealybugs			
Planococcus citri (Risso)	Citrus mealy bug	Citrus, grape	Worldwide
Dysmicoccus bevipes (Ckll.)	Pineapple mealy bug	Pineapple	Africa, Asia, Australia, Pacific Islands, South America
Moths			
Cryptophlebia leucotetra (Meyr.)	False codling moth	Citrus, avocado, stone fruit, guava	Africa
Cydia pomonella (L.)	Codling moth	Apple, pear, peach, quince, *Prunus* spp., walnut	Worldwide
Maruca testulalis (Geyer)	Bean pod borer, mung moth	Legumes	Africa, Asia, Australia, Central and South America, Pacific Islands
Lobesia botrana (Schiff.)	Vine moth	Grape	Europe, Japan, Africa
Scale insects			
Aonidiella aurantii (Maskell)	Red scale	Citrus	Worldwide
Lepidosaphes beckii (Newm.)	Purple scale	Citrus	Worldwide
Quadraspidiotus perniciosus (Comst.)	San José scale	Deciduous fruits	Worldwide
Weevils			
Cylas formicarius (Fab.)	Sweet potato weevil	Sweet potato	Africa, Asia, Pacific Islands, North and South America
Graphognathus leucoloma (Boh.)	White fringed weevil	Root vegetables	South Africa, Australia, New Zealand, USA, South America
Sternochaetus mangiferae (Fab.)	Mango seed weevil	Mango	Africa, Asia, Australia

* Prepared from *Distribution maps of insect pests* issued by the Commonwealth Institute of Entomology, London.

Disinfestation protocols employing either chemical (e.g. hydrogen cyanide, carbon disulphide, methyl bromide [MB], phosphine) or physical (e.g. low temperature, vapour heat, irradiation) treatments have been developed to kill insects infesting horticultural produce in order to meet quarantine requirements. Fumigation with gaseous sterilants is the most effective technique for disinfesting produce. However, these are increasingly unpopular (or banned) because of high mammalian toxicity (e.g. hydrogen cyanide), flammability (e.g. carbon disulphide) and damage to the atmospheric ozone layer (e.g. MB). The manufacture of MB was intended to cease in the USA from the year 2000. However, as no equally effective replacement has been found for postharvest treatment of fresh produce and cut flowers, permission to use it continues. Currently, MB is most widely used for fresh produce, and phosphine for dry or stored produce. High carbon dioxide and/or low oxygen atmospheres have been investigated as alternatives to chemical fumigants, but with limited success. However, high carbon dioxide concentrations are used commercially to control insects infesting dried grapes.

Maximum permissible residues of chemical disinfestants are specified by law in most countries. Subject to the requirements of regulatory authorities in each country, fumigants can be applied to produce in a permanent fumigation chamber, in a temporary enclosure (e.g. under a tarpaulin) or in gas-tight rail cars, road trucks and sea shipping containers. Fumigation under tarpaulins offers flexibility at the dockside or railway yard, and if correctly carried out is as effective and safe as chamber fumigation.

Aerosol formulations of insecticides, such as synthetic pyrethrins or dichlorvos, are used to disinfest ornamentals, such as cut flowers. However, unlike gases, aerosols comprise small droplets that settle out of the air onto surfaces. Consequently, insects that are shielded from the aerosol droplets by an overhead canopy of stems, foliage or flowers, and by the impenetrable walls of plastic buckets, are often not harmed. Postharvest insecticide sprays and dips are, however, often used for cut flowers and foliage and for pot plants, as well as for some fruit and vegetables. The insecticide dichlorvos also has limited vapour phase activity. Slow-release dichlorvos-based pest strips have been included in cartons packed with flowers to effect ongoing disinfestation during export.

Minimum effective doses of chemical or physical disinfestation treatments that leave no survivors have been established through research for a range of insects and horticultural commodities (Plate 11). For produce sensitive to chemical disinfestants, alternative disinfestation procedures have been developed that mainly involve high or low temperatures. The commercial use of temperature treatments has also increased because of concern by authorities and consumers about chemical residues in produce.

Many insects that infest produce cannot tolerate prolonged exposure to low temperatures. This has led to an effective disinfestation procedure for deciduous fruits and some citrus fruit, but not for tropical and subtropical fruits, which are liable to chilling injury. Storage at <1.6°C for 16 days has been shown to be effective for disinfesting fruit against Mediterranean and Queensland fruit flies. Some countries require cold disinfestation to be applied before export, while others permit in-transit cold treatment. Cold treatments may also be applied in combination with chemical fumigation, thus reducing the amount of fumigant required.

Produce can also be successfully disinfested by exposure to high temperature. High temperature treatments can be achieved with hot air, including vapour heat (Figure 11.1), or hot water. For instance, with vapour heat treatment, the temperature of fruit such as citrus and mango and some vegetables is raised to about 47°C using air saturated with water vapour and the core temperature of the fruit is maintained at that elevated temperature for a prescribed period of time (e.g. 15 minutes). As with low temperature treatments, the precise high-temperature treatment regimes, which must be strictly complied with, are specific to the commodity, pest and method; these are the subject of negotiated agreements between the exporting and importing countries.

Irradiation

Ionising radiation is regarded as a mature technology and a safe procedure for killing microorganisms, insects and parasites to improve the safety and quality of many foods and food products. Approval for the use of this technology is, however, country and product specific. The revised Codex Alimentarius General Standard for Irradiated Foods 2003 provides that 'the maximum absorbed dose delivered to a food shall not exceed 10 kGy except when necessary to achieve a legitimate technological purpose'.

Potential benefits of gamma, electron beam or X-ray radiation in the

Figure 11.1 Commercial vapour heat treatment unit used to disinfest mango fruit

(Used with permission of Swift and Co, Thailand.)

postharvest handling of fruit and vegetables include both insect and disease disinfestation, and retardation of aspects of produce development, such as ripening and sprouting (Table 11.2). While these potential benefits have been recognised for more than 30 years, adoption of the technology has been slow, due mainly to public perceptions and acceptance. However, this may change with the introduction of more and tighter restrictions on disinfestation using chemicals, including clear labelling on the products.

Table 11.2 Comparison of maximum tolerable and minimum radiation dose required for desired technical effects on selected fresh produce

Produce	Desired technical effect	Max. tolerable dose (kGy)*	Min. dose required (kGy)
Apple	Control scald and brown core	1–1.5	1.5
Apricot, peach, nectarine	Inhibit brown rot	0.5–1	2
Asparagus	Inhibit growth	0.15	0.05–0.1
Avocado	Inhibit ripening and rot	0.25	–
Banana	Inhibit ripening	0.5	0.30–0.35
Lemon	Inhibit *Penicillium* rots	0.25	1.5–2
Mushroom	Inhibit stem growth and cap opening	1	2
Orange	Inhibit *Penicillium* rots	2	2
Papaya	Disinfest fruit fly	0.75–1	0.25
Pear	Inhibit ripening	1	0.25
Potato	Inhibit sprouting	0.2	0.08–0.15
Strawberry	Inhibit grey mould	2	2
Table grape	Inhibit grey mould	0.25–0.50	–
Tomato	Inhibit *Alternaria* rot	1–1.5	3

* 1 Gray = 100 rads.

SOURCE Adapted from E.C. Maxie, N.F. Sommer and F.G. Mitchell (1971) Infeasibility of irradiating fresh fruits and vegetables. *HortScience* 6, pp. 292–94.

Disinfestation of cold-sensitive tropical fruit (e.g. papaya) by ionising radiation is technically feasible. The egg phase of the life cycle of insects is the most sensitive to irradiation, followed in order by larval, pupal and adult stages. Most insects are sterilised at doses of 0.1–1 kGy. However, some adult moths will survive 1 kGy, although their progeny are sterile. Irradiation is being used in Australia to produce sterile male Queensland fruit flies for release in horticultural production districts to eradicate low-level infestations. Installation of an X-ray irradiator in Hawaii has enabled growers and packers to export papaya fruit to mainland USA. Irradiation at doses of 0.05–0.3 kGy is an effective alternative to chemical treatments for preventing sprouting in potatoes, onions, garlic and yams. This dose range has little effect on other aspects of potato and onion quality, such as sugar level, rates of decay and water loss, texture and flavour. Irradiation of potato and onion is more expensive than treatment with the chemical sprout inhibitors CIPC and maleic hydrazide, but is residue-free. Doses from 0.025–0.75 kGy have been shown to delay ripening in avocado, banana, some cultivars of mango, and papaya. Considerably higher doses of 1–3 kGy are required to delay mould growth in fruit.

Superficial scald control (apples)

The superficial scald disorder of apple in cool storage is associated with the oxidation products of α-farnesene. These products arise when the natural antioxidants in the fruit are degraded or inactivated during cool storage (see Chapter 8). The addition of various synthetic antioxidants effectively prevents the oxidation of α-farnesene and hence the development of scald. Commercially, scald has been reduced by a postharvest dip in either diphenylamine (0.1–0.25%) or 1.2-dihydro-6-ethoxy-2.24-trimethylquinoline (ethoxyquin) (0.2–0.5%). Diphenylamine may also be applied by means of impregnated wraps or in wax and both antioxidants may be brushed onto fruit. Most countries have approved one of the compounds, but rarely both. This has caused problems in the international trade in apples, where exporting countries have to use the scald treatment that is legally permitted by the importing nation. As the treatment is applied immediately after harvest, the final destination of the fruit must be known with some certainty at harvest, and some segregation of treated fruit must be maintained during storage.

Although diphenylamine and ethoxyquin adequately control scald, a single compound, approved by all countries, would be desirable. For political reasons, it is unlikely that either compound will gain universal approval. Chemical companies are reluctant to pursue approval for various other compounds that have been shown to be effective scald inhibitors due to the increasingly stringent demands of health authorities. However, application of 1-MCP before cool storage has been shown to control scald. Ethylene production is associated with the synthesis of α-farnesene and 1-MCP blocks this action. 1-MCP has been recently approved as a treatment for apples in many countries.

There is a reluctance internationally to allow new chemicals as food treatments. This concern particularly affects the postharvest treatment of fresh produce, as food laws are mainly concerned with what happens after harvest. In addition, strict MRLs have been established for agricultural pesticides and fungicides. Research with superficial scald has focused on finding volatile compounds, such as ethanol, that are naturally produced by fruit and vegetables as possible commercial treatments that may have universal acceptance. While the addition of ethanol vapour during cool storage will control scald, stringent control of ethanol use in some countries may make it difficult to gain approval.

Calcium application (apples)

Since about 1960, preharvest calcium sprays have been applied commercially to apple in several countries to reduce incidence of the physiological disorder bitter pit (dark necrotic spots on the skin) after harvest (see Chapter 8). Added calcium also reduces internal breakdown of apple in cool storage. To be effective, calcium must actually contact the fruit and be absorbed directly by it. Calcium that falls on the leaves or branches is not effectively transferred to fruit. The sprays may be applied up to six times during the growing season in order to slowly build up calcium levels in fruit. It is difficult to ensure that all fruit are wetted sufficiently by one application of calcium solution. The preharvest spray uptake problem may be overcome by dipping apples after harvest in solutions of calcium salts. Uptake can be further improved by partial pressure (reduced pressure) or positive pressure infiltration to force calcium solution into the apple flesh. The best results are obtained with apples that have a closed calyx so that the calcium solution is forced into the fruit through the lenticels and is thus spread around the perimeter tissue where the disorders occur. With open-calyx fruit, the uptake of solution is difficult to control as it readily enters the fruit via the calyx and excess solution accumulates in the core area, often leading to injury or rotting.

Laboratory studies on partial-pressure infiltration of calcium solutions have shown that the technique can retard the initiation of ripening in a number of climacteric fruit, such as tomato, avocado, mango and pear. However, large-scale application of calcium infiltration into fruit other than apple has not been undertaken due to the attendant risks of skin injury due to excess calcium uptake and rot development.

Controlled ripening

Climacteric fruits are frequently harvested green-mature or earlier than fully ripe (e.g. banana, European pears, kiwifruit, mango) and then transported, often over considerable distances, to areas of consumption. These fruits are then ripened to optimum quality under controlled conditions of temperature and RH, and, with some fruits, through the addition of ethylene and other gases that initiate ripening. A further advantage of controlled ripening is improved uniformity of ripening among fruit. The use of relatively high ripening temperatures may also minimise the development of rots in ripe tropical fruits. In contrast, non-climacteric fruits undergo little or no desirable change in composition after harvest, and are not harvested until they are fit for consumption.

A significant proportion of the world production of banana is ripened under controlled conditions. The banana is unusual in that it can be picked over a wide range of physiological maturity ages, from half-grown (thin and angular) to fully grown (full and rounded) and ripened to high quality with the aid of ethylene. Ethylene is by far the most active of the known ripening gases, which are all low-molecular-weight hydrocarbons. Acetylene, generated by adding water to calcium carbide, also induces the ripening response; but in practice a concentration at least 100 times higher than ethylene is required. The commercial ripening of bananas is a routine operation and fruit at a specified colour stage (Table 11.3) can be produced on a predetermined schedule. The effective concentration of ethylene for banana ripening is quite low (Table 11.4) and when this concentration is maintained for the stated period, further increases in concentration give no added advantage. In practice, however, high concentrations are used initially because the ripening rooms are often not sufficiently airtight. Bananas may be ripened by a batch process in which the chamber (Figure 11.2) is charged (the 'shot') with ethylene gas to a concentration of 20–200 µL/L. The chamber has to be ventilated after the first 24 hours to prevent the accumulation of carbon dioxide. Carbon dioxide concentrations should not exceed 5000 µL/L (0.5%) to allow personnel to enter the rooms to inspect the fruit. If the chamber is poorly sealed, it may be necessary to recharge with ethylene after 12 hours. A more satisfactory alternative to charging the ripening chamber with an initial high concentration of ethylene is to 'trickle' ethylene into the chamber (Figure 11.3) at a rate just sufficient to maintain the concentration shown in Table 11.4. The ripening chambers should be ventilated at the rate of about one room volume each 6 hours, to prevent the accumulation of carbon dioxide. In practice it is usually not necessary to install a ventilation system in rooms less than 60 m^3 because they have natural air exchange (leakage) rates higher than the required minimum rate. For larger rooms, the natural air leakage rates should be taken into account when calculating the ventilation requirement. Excess ventilation will increase energy consumption for cooling and heating.

Traditionally, containers of bananas were stacked in a ripening room so

Table 11.3 Colour stages of the ripening Cavendish banana

Stage	Peel colour	Approx. starch (%)	Approx. sugar (%)	Comments
1	Green	20	0.5	Hard, rigid; no ripening
Sprung	Green	19.5	1.0	Bends slightly; ripening started
2	Green, trace of yellow	18	2.5	
3	More green than yellow	16	4.5	
4	More yellow than green	13	7.5	
5	Yellow, green tip	7	13.5	
6	Full yellow	2.5	18	Peels readily; firm ripe
7	Yellow, lightly flecked with brown	1.5	19	Fully ripe; aromatic
8	Yellow with increasing brown areas	1.0	19	Over-ripe; pulp soft and darkening, highly aromatic

SOURCE Commonwealth Scientific and Industrial Research Organization (1972) Banana ripening guide, CSIRO, Melbourne, Division of Food Research, circular 8.

Figure 11.2 A modern double-deck controlled ripening room for bananas using forced-air circulation. The boxes are vented to allow air to be drawn through the packed fruit.

as to expose at least two faces to the circulating air, ensuring that fruit temperatures were even. In modern practice, the fruit is packed in vented cartons, unitised on pallets, and fruit temperatures are controlled by forced-air circulation. Adopting the forced-air system has increased the room capacity by about 40 per cent and enabled fruit temperatures to be controlled more accurately. A minimum air-flow of about 0.34 L/ sec.kg of bananas is required.

Dessert cultivars of peaches, nectarines and some Japanese-type plums that retain relatively firm texture when ripe can be conditioned or control ripened. These fruit are usually

Table 11.4 Ripening conditions for some fruits using ethylene*

Fruit	Temperature (°C)	Ethylene (μL/L)	Treatment time (hours)
Avocado	18–21	10	24–72
Banana	15–21	10	24
Cantaloupe	18–21	Nil	n/a
Honeydew melon	18–21	10	24
Kiwifruit	18–21	10	24
Mango	29–31	10	24
Papaya	21–27	Nil	n/a
Pear	15–18	10	24
Persimmon	18–21	10	24
Tomato	13–22	10	continuous

* RH is normally maintained at 85–90%.

Figure 11.3 Trickle system for adding ethylene to a ripening room. (1) ethylene cylinder; (2) pressure regulator with outlet pressure set on 300 kPa; (3) on–off toggle valve; (4) solenoid valve wired to open only when the air circulation fan is operating; (5) variable flow controller or rotameter located on the front wall of the room; (6) air circulation fan or refrigeration unit; (7) gas outlet located near the centre of the room in small rooms (less than 60 m^3); (8) in larger rooms, the gas outlet is located in front of the outlet from the blower fan; (9) blower fan mounted on the front wall of the room (the amount of air discharged into the room can be regulated to provide a predetermined rate of ventilation); (10) branches to other rooms if required; (11) a small exhaust porthole is provided in the rear wall of rooms (usually large rooms over 60 m^3) that require mechanical ventilation.

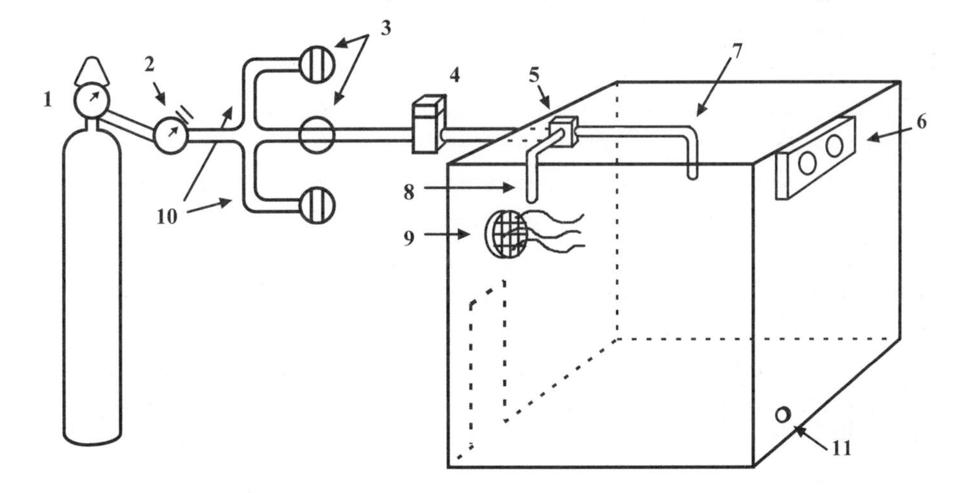

SOURCE W.B. McGlasson, E.E. Kavanagh and B.B. Beattie (1986) Ripening tomatoes with ethylene. *Agfact*, H8.4.6, NSW Department of Agriculture, Sydney.

harvested at a maturity stage several days prior to the eating ripe stage to minimise bruising during grading and packing. Packed fruit can be ripened to within about one day of eating ripe at 20°C and high RH using forced-air circulation. Once the desired stage of ripeness, based on firmness measurements, is reached the fruit must be rapidly cooled by forced air to <2°C and maintained at low temperature until placed on retail display. Conditioning for up to 48 hours may delay the development of low-temperature disorders in some cultivars.

Water loss can be high at ripening temperatures unless RH is maintained at a high level. Humidity can be raised by atomising water into the ripening chamber. Regulating RH during the course of ripening can be particularly important for banana. A RH range of 85–90% has been recommended for ripening to stage 2 (Table 11.3), but this should be reduced to 70–75% during the later colouring stages to avoid skin splitting. Although the best skin colour may be achieved at the highest RH, commercial experience with conventional chambers not designed for forced-air circulation has shown that the skin tends to be too soft and may split. If RH is too low, weight loss may be excessive, colour poorer and blemishes more pronounced. Maintaining RH at 85–90% continuously in forced-air ripening rooms has proved satisfactory. Because of the high RH and temperature maintained in ripening rooms, moulds grow readily on any organic matter, including on the walls of the room if they are not suitably protected. Regular cleaning with a solution of sodium hypochlorite (chlorine) is required. Ozone generators can also keep surfaces clean, but must be used correctly so as not to expose people to toxic concentrations (Chapter 7).

The ripening of other climacteric fruit that has been harvested immature can be hastened by treating it with ethylene under controlled conditions. However, in contrast to banana and avocado, quality will be inferior to that of fruits harvested at the mature-green stage. With many climacteric fruits, it is important to harvest at the correct stage of maturity. For example, at the full-slip stage for cantaloupe, the first appearance of yellow colour in the blossom end of papaya, and the first colour (breaker) stage of tomato. However, at full maturity, it is only necessary to hold fruit at the temperature and RH specified in Table 11.4 to achieve high-quality ripened fruit. Thus, treatment with ethylene is not necessary for these fruits to ripen fully. When fully developed, at least some (if not all) of the fruit will produce sufficient ethylene to effectively ripen itself and adjacent fruit. Nonetheless, ethylene treatment will promote more uniform and rapid ripening in consignments of mixed maturities (e.g. once-over [strip] picked mango). In Australia, forced-air systems are also common in tomato ripening rooms.

Controlled degreening

The pulp of many early-season citrus cultivars becomes edible before the green colour has completely disappeared from the peel. Exposure to low temperature during maturation is necessary for an orange-coloured peel to develop. This requirement explains why the peel of citrus grown in the low-altitude tropics fails to degreen completely. Furthermore, the Valencia orange cultivar is often stored on the tree for several months after ripening has been completed, and during this storage period the peel tends to regreen. Postharvest treatment with ethylene under controlled conditions hastens the loss of chlorophyll, a process known as degreening. Batch or trickle degreening is a cosmetic treatment designed to give fruit a ripe appearance, but it does not result in significant changes in pulp composition if correctly administered. The conditions of batch degreening, 20–200 μL ethylene/L, 25–30°C and 90–95% RH, are maintained for 2–3 days with regular ventilation of the chamber to prevent carbon dioxide build-up (citrus is injured by carbon dioxide concentrations above 1%). Trickle degreening, with 10 μL ethylene/L continuously metered into the room, is more rapid than batch degreening and is therefore preferable, since degreening conditions accelerate deterioration and decay of citrus. Although the most rapid degreening occurs at 25–30°C, production of peel carotenoids is greatest at 15–25°C. Fruit may be ripened and degreened equally well using the ethylene-releasing compound ethephon. Ethephon is absorbed by fruit tissues and, when the pH exceeds 4.6, breaks down to release ethylene.

In some citrus-growing areas, notably Florida in the USA, tangelos, temples and some early-season oranges do not experience enough cold weather to promote the development of a highly coloured peel. Packing houses in these areas have been legally permitted to dye the peel of these fruits, under strictly controlled conditions, with Citrus Red No. 2. This process is known as 'Colour Add', and can only be used on mature fruits that are not intended for processing. The dye is applied to citrus fruit by dip or drench, and typical treatment times and temperatures are: 4 minutes (max.) at 49°C for oranges and 4 minutes (max.) at 46°C for temples and tangelos. After treatment, the fruit are rinsed thoroughly to prevent 'bleeding' of the dye through the wax and to ensure that the Food and Drug Administration (FDA) residue tolerance of 2 mg/kg fruit is not exceeded. Wax is applied after all water is removed.

Light (minimal) processing

There is strong consumer demand for ready-to-use foods, including fruits and vegetables. This demand has fostered the development of the 'lightly processed' fruit and vegetable industry, which is also known as the 'minimally processed' or 'fresh cut' industry (Plate 12). The generalised process for vegetables involves carefully scheduled production, early morning harvest, immediate fast cooling (e.g. by vacuum cooling), refrigerated transport, trimming, cutting, washing (e.g. with potable water), disinfecting (e.g. with chlorinated water), de-watering (e.g. by centrifugation), mixing (e.g. of green and red cabbage), packaging (e.g. vacuum or pillow packs; gas flush with a modified atmosphere), and distribution and sales at low temperature (e.g. 5°C). Fruit slices (e.g. mango 'cheeks') or pieces (e.g. pomelo segments) are also utilised. The modified gas flush atmospheres (low oxygen, high carbon dioxide) minimise oxidative browning, especially at cut surfaces. Important process variables include methods of cutting (e.g. knives, lasers), equipment maintenance (e.g. knife sharpening) and angles of cut, since such variables determine the degree of tissue wounding. Good hygiene goes hand in hand with low-temperature handling, these being critical precautions against growth of potentially toxic micro-organisms (e.g. *Listeria* spp.).

An extension of the lightly processed scenario is fruit drying, which has long been practised for grapes (e.g. sultanas), but is also applied to many other temperate, sub-tropical and tropical crops (e.g. mango fruit leather, sun-dried tomatoes). High temperature and low RH air processing conditions minimise browning reactions, which might otherwise detract from produce appearance.

Pulsing

Some fresh cut flowers are 'pulsed', whereby a solution is supplied via the transpiration stream. In the most basic instance, the chemical is water and the process is hydration. Hydration may be facilitated by adding a wetting agent to the water. Cut flowers may also be pulsed with a sugar, such as sucrose. Sucrose pulsing usually involves concentrated sucrose solutions (e.g. 5–20% w/v) supplied for a matter of hours (e.g. overnight) at room temperature or in a cool room. Sucrose pulsing of gladiolus supplies an osmoticum and a respirable substrate to sustain subsequent bud opening in the vase. Cut flowers such as carnations and delphiniums are pulsed with anti-ethylene agents such as STS or AOA. AOA is toxic to most flowers, except carnations. Typical STS pulsing protocols are 2–4 mM silver ions (as the STS complex) for 15–45 minutes at ambient temperatures or 0.5 mM silver overnight at about 1°C. Cut flowers are also pulsed with dyes, such as the food-grade blue dyes used on white carnations to give interesting visual effects (e.g. blue-coloured petal veins and margins).

Cut flowers and foliage destined for desiccation can be pulsed for one to a few days with a humectant, such as 20–30 per cent v/v glycerol. This process is known as uptake preservation. Instead of desiccating completely, plant material treated in this way retains a degree of suppleness (plasticity), associated with the humectant chemical attracting water vapour from the atmosphere into the tissue. As drying plant tissue, particularly when pulsed with a humectant, often browns, red, green, blue and other coloured dyes are frequently supplied along with the humectant.

Grading

Grading is a critically important process because produce presentation, an aspect of quality, is often judged on the basis of uniformity. Uniformity is important in terms of presenting a standard product for handling (e.g. unitisation) and marketing. Produce is generally graded on the basis of size, weight, colour, defects or composition; or a combination of these features. A typical size grader for fruit separates produce using diverging conveyor belts, which allow smaller fruit to drop between them first. Computerised weight graders can operate on the basis of tipping buckets that drop to release the preweighed item at a particular position. Video image capture and analysis can also be used for size grading, as well as for colour grading and external defect grading. Video imaging and analysis technology is potentially important for ornamentals, such as cut flowers and pot plants. Composition is not yet widely used as a means of grading fruit. However, automated systems using NIR spectrometers to assess soluble solids concentration non-destructively in fruit such as apples and stone fruit have been developed and are in commercial use in some parts of the world. Technologies with potential application in grading include X-ray imaging and computer-aided tomography, proton or chemical shift magnetic resonance imaging, transmitted and reflected light spectroscopy, acoustic response signals to tapping, and volatile emissions analysis using solid state detectors (Chapter 10). However, such advanced technologies are not likely to find practical application until capital cost is minimised and throughput (items/sec.) is maximised. There is an ever-present need to grade out sub-standard produce on the basis of internal defects, such as jelly seed in mango and light brown apple moth in apple; and eating quality, such as acceptable sweetness and flavour in honeydew melon – but these applications are technically much more complex.

12 Packaging

Centres of consumption of fresh produce are usually remote from production areas. Accordingly, the allied costs of distribution, which include handling, packaging and transportation, often exceed those of production in both money and energy terms. Careful management of the distribution system will ensure that produce retains its quality and that economic returns are maximised. Packaging has been practised for as long as fresh produce has been traded. Its two main functions are to assemble the produce into convenient units for handling (unitisation), and to protect the produce during distribution, storage and marketing.

The earliest packages were mostly constructed of plant materials, such as woven leaves, reeds and grass stems (Figure 12.1), and were designed to be carried by hand. Even today, most packages are handled manually at some stage, and are sized accordingly. Nevertheless, packages are often assembled into larger units for mechanical handling by forklift (e.g. pallet loads) or crane (e.g. sea, road and rail containers). Nowadays, produce is transported and sold in a wide range of packages, constructed of wood, fibreboard, jute (hessian) or plastics (Figure 12.2 and Plate 13). Many packaging systems waste raw materials, but environmental and food safety concerns have led to recycling fibreboard and using returnable plastic boxes that are washed and reused many times (Figure 12.3).

Modern packages and packaging for fresh produce are expected to meet a range of basic requirements. They should:

Figure 12.1 Typical bamboo baskets used throughout South-East Asia for handling and transporting of produce

Figure 12.2 Bulk bins made from fibreboard

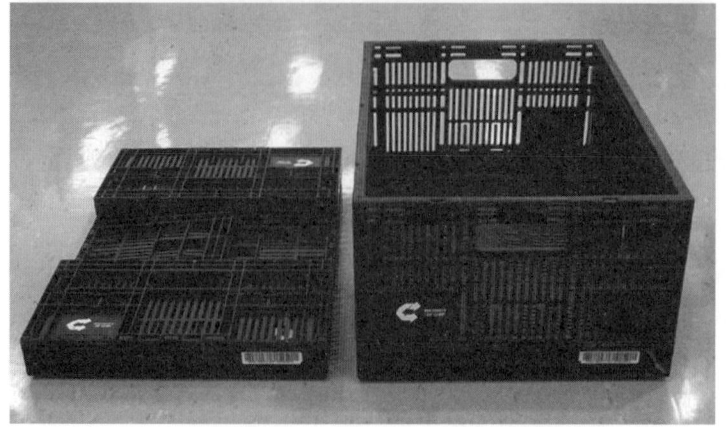

Figure 12.3 Returnable collapsible plastic boxes used for 'retail ready' display of produce

- have sufficient mechanical strength to protect the contents during handling and transport, and while stacked.
- be largely unaffected, in terms of mechanical strength, by moisture content when wet or at high RH.
- stabilise and secure product against movement within the package during handling.
- not contain chemicals that could transfer to the produce and taint it or be toxic to the produce or to humans.
- meet handling and marketing requirements in terms of weight, size and shape.
- allow rapid cooling of the contents, and/or offer a degree of insulation from external heat or cold.
- utilise gas barriers (e.g. plastic films) with sufficient permeability to respiratory gases as to avoid any risk of anaerobiosis.
- offer security for the contents, and/or ease of opening and closing in some marketing situations.
- identify the contents, proffer handling instructions and aid retail presentation through comprehensive and accurate labelling.
- either exclude light (e.g. from potato) or be transparent (e.g. for orchids).
- facilitate easy disposal, reuse or recycling.
- be cost-effective in relation to the value and the required extent of protection of the contents.

The need to reduce the many types (sizes and shapes) of packages through standardisation as a means of reducing waste materials and costs has been recognised by large retail chains. This has led to adoption of fibreboard trays and reusable plastic boxes with a standard 'foot print' that fully utilise the area of the standard pallet, especially for fresh fruit. The packed fruit is often placed directly on retail display (retail ready), so retail staff no longer need to hand-stack displays and there is less handling of the fruit by consumers. Unitisation (e.g. pallets) and mechanical handling (e.g. forklifts) make standardisation essential for economical operation.

Prevention of mechanical damage

Fruit and vegetables vary widely in their susceptibility to mechanical damage and in the types of mechanical injury to which they are susceptible. Accordingly, the choice of package and packing method must take these differences into account. Four different causes of mechanical injury to produce can be identified: vibration, compression, impact and cut. The relative susceptibilities of some fruits to compression, impact and vibration injuries are shown in Table 12.1.

Table 12.1 Susceptibility of produce to types of mechanical injury

Produce		Type of injury		
		Compression	Impact	Vibration
Apple		S	S	I
Apricot		I	I	S
Banana	green	I	I	S
	ripe	S	S	S
Cantaloupe		S	I	I
Grape		R	I	S
Peach		S	S	S
Pear		R	I	S
Plum		R	R	S
Strawberry		S	I	R
Tomato	green	S	I	I
	pink	S	S	I

S = susceptible; I = intermediate; R = resistant

SOURCE R. Guillou (1964) *Orderly development of produce containers. Proceedings: Fruit and Vegetable Perishables Handling Conference*, 23–25 March. University of California, Davis CA, USA, pp. 20–25.

With the possible exception of certain so-called 'hard' vegetables, such as watermelon, pumpkin, onion, carrot and potato, the package must be strong enough to carry stacking loads – otherwise there will be compression bruising of the contents. Impact bruising is caused by dropping packages and by other impact shocks during handling (e.g. on the grading line) and transport. Bruising is tissue damage that results from strain energy being dissipated in the tissue. The amount of damage depends on how much energy is dissipated and the nature of the tissue. Vibration injury is common during transport, resulting in abrasion marks ranging from light rubbing to removal of the skin and possibly some of the flesh. The problem of cuts (e.g. sharp edges) and punctures (e.g. nails, stems) also merits attention.

Nearly all mechanically damaged tissue turns brown through enzymatic (e.g. polyphenol oxidase) or chemical (atmospheric oxygen) oxidation of phenols. Tissue deformation and browning is exacerbated by increased water loss as a result of damage to the cuticle. Furthermore, injuries represent sites for microbial infection in terms of both damage to the protective skin and release of substrates for growth (e.g. sugars). The physiological wound response involves increased ethylene production and respiration, and hence acceleration in the rate of produce deterioration. Mechanical damage also causes an immediate loss of edible material, because of the need to trim off unsightly portions or discard whole units.

Two important practical requirements must be met when packaging perishable produce:

1. Individual items should not be allowed to move with respect to each other or to the walls of the package, in order to avoid vibration injury.
2. The package should be full without being packed too tightly, which increases compression and impact bruising.

In short, both underfilling and overfilling are to be avoided. Individual items should be held firmly, but not too tightly, within the package. Packaging can be made more protective by individually wrapping each piece of produce (e.g. paper wraps), by isolating each piece (e.g. cell and tray packs), or by using energy-absorbing materials (e.g. cushioning pads). However, such additional packaging increases cost, and must therefore be justified by reduced wastage, increased selling price, or a reputation for reliable quality such that price and sales are maintained during periods of over-supply. Careful handling of packages is the best general precaution against mechanical damage.

The potential for vibration damage during transport deserves particular emphasis. Each product starts to vibrate (move around; rotate) at a certain product-specific excitation frequency (Hz). In the case of packaged product in stacks (e.g. palletised) during road transport, the vibrations resulting from the interaction of the road surface and the vehicle suspension system (2–20 Hz) will also be propagated with increasing magnitude up through the stack. For these reasons displaced cartons and produce affected by vibration injury are often seen at the top of stacks. Various strapping techniques are commonly used to stabilise the stacks.

Cooling produce in the package

A further important requirement of packaging is that it is usually required to allow rapid cooling of produce. For example, containers designed for pressure cooling (Chapter 4) should have holes occupying about 5 per cent of the surface area on each of the air entry and exit ends. Both the nature of the produce and the treatment after packing must be considered in designing containers for specific commodities. Ideally, respiratory heat should be able to escape readily from packages. In the case of small and/or tightly packed commodities, such as green beans, small fruits, leafy vegetables and cut flowers, the vital heat of respiration is removed largely by conduction to the surface of the package. Thus, the mass of the contents (i.e. the minimum dimension of the package from the centre to the surface) becomes a critical issue. The acceptable mass depends on the respiration rate of the commodity. If the mass of produce is excessive, that near the centre of the package will heat up because respiratory heat cannot dissipate fast enough. In practice, self-heating is a significant problem for rapidly respiring produce such as peas, lettuce and broccoli in large single packages or in close stacks of packages. The problem may be avoided with smaller packages or by ventilating large packages or stacks.

Bulk bins that hold 200–500 kg of produce are now widely used for harvesting, storage and often transport of fruit and vegetables. Adequate cooling of fruit in stacks of bins may be achieved in cool stores without forced movement of air, through a 'chimney effect', if about 10 per cent of the floor area of these bins is vented. Nevertheless, fan-assisted ventilation is necessary for rapid overnight cooling.

Effects of packaging on weight loss

Packaging is often used to minimise weight loss and shrinkage of produce during marketing. Specific approaches include wrapping individual items with waxed paper or plastic film wrapping, bagging individual items, consumer packaging of a number of items in trays (e.g. polystyrene) over-wrapped with a moisture barrier material, and increasing the moisture resistance of fibreboard packages with a surface or internal plastic laminate or by lightly or fully waxing the carton.

Under dry conditions, containers (e.g. wooden boxes, plastic crates, baskets) of produce may be deliberately sprayed with water. Direct wetting can also assist in cooling produce by evaporating the added water. Fresh cut flowers and foliage are often transported 'wet', usually in plastic buckets, but sometimes with stems in individual phials of solution.

Package dimensions

Package dimensions are both economically and structurally important. The size and shape should facilitate economic use of materials. Packages should also offer adequate strength and allow easy and secure handling, loading and stacking. An optimal length to width ratio is about 1.5:1. There is a trend towards smaller packages because of recommendations by the International Labour Organization (ILO) concerning the maximum weights people can reasonably be expected to handle routinely. Thirty litre (about 20 kg of produce) and 15 L packages are becoming standard for fruit, with a larger 36 L package being used for some vegetables. Retail-ready packages of fruit may contain only 5 kg of fruit in a single layer.

Standardisation of package sizes promotes efficient handling. Package sizes are standardised by determining the dimensions and numbers that fit well on standard pallets. However, achieving this objective may require that produce be 'grown to fit the package' rather than the package being sized to fit the produce, as in the past. Traditional baskets and hessian or plastic mesh bags, although relatively cheap, do not stack efficiently. Also, they are often too large for efficient handling and there is a tendency to over-pack them with produce (Figure 12.1). Produce therein is often damaged during handling and transport, because of the lack of structural support to protect the contents.

Mechanical strength of packaging

For continued protection of produce against mechanical damage, packages must retain their strength throughout the marketing chain. A common saying is that: 'the package should support the produce, rather than the produce support the package'. Wood and solid and expanded plastic packages are inherently strong, compared to fibreboard packages. However, wood is an expensive and environmentally costly material. Solid plastic containers are even more expensive, although they are amenable to washing and reuse. Versions that can be collapsed or folded-down and reused are gaining acceptance because they require only a small amount of space for the return journey from retail store to distributor. Rigid expanded polystyrene is lightweight yet strong, but it requires considerable storage space and must be recycled using expensive capital infrastructure into high-density material. In comparison, fibreboard is attractive and can be made stronger by using two or more thicknesses, such as the bottom and lid of fully telescoping cartons. The strength of fibreboard lies in the fluting between the inner and outer liners. Fibreboard comprised of two layers of fluting sandwiched between three liner layers is stronger than the conventional single layer of fluting.

Under conditions of high humidity, after condensation or after being wet by rain, the strength of the package must either be independent of moisture content (e.g. wood) or the package material must not absorb moisture (e.g. plastic). Commonly used fibreboard cartons and trays rapidly lose strength as they absorb moisture and so are less than satisfactory under tropical conditions and in high-humidity cool storage. Fibreboard can be protected if fully impregnated with wax or a similar material, but wax impregnation is expensive and waxed fibreboard is not recyclable. The corrugated fibreboard used in cartons is often only given a surface coating ('lightly waxed'), which affords only a degree of protection from free water or high RH. Packages and/or packing materials might also be required to exclude water from the produce (e.g. to prevent grape splitting) or to prevent dehydration of the produce (e.g. roses, which otherwise suffer 'bent neck').

Packing

Tightly filled packs are desirable for most fruits and vegetables, but without underfilling or overfilling. The package, and not the produce, should bear the stacking load. However, it is equally important that produce does not move and sustain vibration injury during handling and transport. Some produce, such as potato and carrot, and some types of citrus fruit (e.g. orange), will withstand reasonable compressive loads. For these, non-rigid packages, such as mesh bags, are satisfactory provided they are handled with due care. In a similar fashion, some vegetables (e.g. asparagus) and cut flowers (e.g. iris, gladiolus) can be packaged in bundles.

In many countries, fruit has traditionally been place- or pattern-packed, such that each piece was put into a specific position by hand. The objectives of place packing were to maximise net weight, maintain a tight pack and present the fruit attractively when the package was opened. However, this time-consuming approach, which requires accurate grading into a range of size-classes, has become very costly. The alternative is volume- or tight-fill packing, whereby fruit are 'poured' into the carton or box. After filling, the pack is vibrated to obtain reasonably tightly packed fruit within. By this method produce (e.g. apple) is packed to a standard weight rather than a standard count. In a volume-fill pack, produce is often kept firmly positioned beneath a pressure pad (e.g. paper pulp in an envelope) over which the lid is securely fastened. Package inserts, such as moulded pulp or plastic trays to isolate individual fruit, are expensive. Nonetheless, they are commonly used for delicate and/or high value products (e.g. mango) and are ideal for retail-ready displays. Individual wrapping still has a place, as does lining packages with plastic or paper in order to reduce vibration damage and/or moisture loss. Very delicate fruit, such as papaya, may be sleeved in thick spongy plastic mesh. Cut flowers are often packaged in paper or wood 'wool' for protection, and flower bunches and potted plants are often sleeved in plastic film.

It is not practical to recommend specific packages for each fruit, vegetable or ornamental, as several types may be satisfactory. The most suitable package depends on many factors, including the region, environmental conditions, length and nature of the market chain, methods of handling and transport, availability and cost of materials, and whether the produce is to be refrigerated.

Stowage

Stowage or stacking of packages should simultaneously ensure stability of the stack and allow adequate air movement for satisfactory cooling. Stowage must also be economical of space and easy to achieve. Stronger packages are needed for long-distance transport and/or for high stacking in cool stores as compared with immediate local marketing. Where packages are normally handled on pallets, the package must fit the pallet. Stack stability is usually best obtained by either cross-stacking or tied-stacking, both of which require the packages to have suitable length to breadth ratios. Tied-stacking can be achieved using straps, nets, plastic wrap and tape.

Consumer pre-packing

Produce purchased at retail outlets was traditionally packaged in paper bags. However, paper bags have been largely replaced by polyethylene film bags, which are cheaper, stronger and usually transparent. Supermarket outlets have led a trend towards marketing pre-packaged fruit and vegetables, whereby produce is pre-weighed and packaged into small units for retail sale. However, consumers are sometimes wary of the quality of pre-packaged items and many still opt to select individual produce items from an open display (Plate 14).

Consumer packaging of produce in small plastic bags or in plastic or paperboard trays over-wrapped with clear film helps restrict weight loss. Such packaging can also provide a modified atmosphere benefit. However, the modified atmosphere effect is somewhat risky and generally not sought. Accordingly, films with low gas permeability are perforated to prevent significant modification of the package atmosphere. The small perforations necessary do not increase water loss appreciably.

Plastic films for packaging produce should ideally have good tensile strength, gas and water permeability, heat sealability, clarity and printability. Manufacturers can develop films to suit most packing specifications. Low-density polyethylene film is most widely used for consumer packs. Polyethylene has good clarity, can be heat-sealed, is flexible over a wide temperature range (–50 to 70°C), and is probably the cheapest film in most countries. Polyethylene is relatively permeable to many volatile compounds and gases, but comparatively impermeable to water vapour. The gas permeability of films can be controlled by varying either the density of the film or its thickness, or, as mentioned above, the film may be perforated.

Modern packaging

With the rapid advance of technology, there is interest in producing specialised packaging. The term 'active packaging' has been coined to describe packaging that offers a level of control over in-package conditions and how they vary with produce (e.g. ethylene production) and environmental (e.g. temperature) factors. An example might be a polymer film that can increase or decrease in permeability to oxygen and carbon dioxide as temperatures rise and fall, respectively.

Water loss in packages can be reduced by the use of entire or micro- (pin-hole size) or macro-perforated films. However, condensation within moisture barriers can become a problem. In film-wrapped produce (e.g. citrus wrapped in heat-shrink film), condensation is generally not a problem because the film is in intimate contact with the fruit and assumes the same temperature as the fruit. In the case of loose-wrapped produce (e.g. cut flowers within a plastic carton liner, or fruit in a consumer pack), condensation can be reduced using simple or elaborate moisture sinks, such as newspaper or salts in spun-bonded polyethylene sachets, respectively. Chemical anti-fogging treatments can be applied to films, and films with relatively high water permeability can be used (e.g. cellophane, polyvinyl chloride). Importantly, these precautions to avoid condensation must also maintain product visibility (e.g. berry fruit in over-wrapped punnets). Water absorbents can be incorporated in packaging to capture and hold the free water resulting initially from condensation, which is followed by droplet formation and finally pooling.

Ethylene can be scrubbed from the package using blocks or sachets of high surface area materials (e.g. florist's foam, aluminium oxide particles) coated with potassium permanganate or using films impregnated with an oxidant, such as tetrazine. Tetrazine is a promising compound, since, unlike potassium permanganate, it is relatively specific for small double-bonded volatiles like ethylene. However, whatever the active ingredient, ethylene scrubbers are generally unlikely to work effectively unless they are well positioned to intercept ethylene. Ethylene is deleterious at ppb levels and is produced by the harvested plant tissue itself. Concentration differences at ppb levels do not constitute gradients of sufficient magnitude to drive ethylene to a point (e.g. sachet) or planar (e.g. film) sink. Thus, provision of ethylene scrubbers within packages can be a waste of money and effort. It is tempting to envisage small ethylene scrubbers with tiny motorised fans that draw through the in-package air in order to lower ethylene concentrations.

Respiratory gas levels (i.e. oxygen, carbon dioxide) can be controlled by the chemical (e.g. polymer type) and physical (e.g. thickness) characteristics of plastic films, as well as by holes in films. Oxygen and carbon dioxide flux through holes is proportionally greater in magnitude than water vapour and ethylene flux because their flux is driven by comparatively large concentration gradients (i.e. percentages versus ppm or ppb). Thus perforated films can be used effectively to reduce water loss while avoiding the risk of anaerobiosis. All other factors being equal, oxygen diffuses somewhat faster in air (e.g. through holes) than carbon dioxide on account of its greater diffusion coefficient. In contrast, all plastic films are relatively more permeable to carbon dioxide than oxygen. Both oxygen and carbon dioxide can be chemically scrubbed from packages. Lack of consistent temperature control during handling and transport, and differences in the temperature quotient for physical gas diffusion across plastic films as compared to those for physiological processes such as respiration increase the likelihood that anaerobic conditions may occur in sealed plastic film packages. Such risks may be minimised using fail safe (e.g. low melting point polymers) or variable aperture devices (e.g. bimetallic strips) to regulate formation and/or size of holes. Advances in the areas of microelectronic, biosensor and polymer science are likely to yield films that actively sense and respond in a controlled way to stimuli, such as increases in temperature.

In the absence of refrigeration, a certain level of temperature control during handling and transport can be achieved with insulation (e.g. polystyrene boxes) and with heat sinks provided inside the packaging (e.g. loose ice or ice packs). Also in the absence of refrigeration, external reflective and/or insulative covers (e.g. thermal blankets) and heat sinks (e.g. dry ice) can assist, or provide an alternative to, in-package temperature control measures.

Environmental issues

Packaging for horticultural produce has enormous environmental implications because of the large quantities of material involved and their eventual disuse. Wooden packaging is biodegradable, as is fibreboard and paper packaging, which is also recyclable. However, these materials are forest products, and management of forest resources is a strong consumer issue. Solid plastic packaging is mostly reusable. Unfortunately, however, returnable plastic crates are relatively expensive, both in terms of initial outlay (production; purchase or hire) and maintenance costs (e.g. washing, backloading). Polystyrene (expanded polymer) and plastic films constitute significant environmental problems, both aesthetically and to wildlife. Nevertheless, polystyrene can be and is being recycled; for instance, into high-density material. Plastic films can also be recycled, although stringent recycling practices are seldom applied at present. Nevertheless, there is little question that, through public pressure, more environmentally sound packaging will come into use. For example, biodegradable packaging films made from materials such as cellulose, starch and proteins are being developed. In time, if not at present, the use of environmentally sound packaging should offer a marketing edge.

13 Commodity storage recommendations

Earlier chapters have considered a range of technologies available to maintain quality and extend postharvest life. Technologies available for some produce, such as certain apple cultivars, will allow a storage life of 12 months. In contrast, for produce such as highly perishable berry fruit and leafy vegetables, the extension in market life may only be a few days. All postharvest interventions, whether from changes in handling, storage or transport systems, come at some cost, which must be recouped from an enhanced market return. It is, therefore, a subjective value judgment as to when and where to market and, thus, what postharvest technology is required.

The principle processes of postharvest deterioration to be inhibited are the general metabolic rate, water loss, and wastage caused by pests and diseases. The main technological interventions are to control the temperature and humidity of the atmosphere around produce. For most commodities, the optimum temperature is just above the freezing point and the optimum RH is just below 100%. However, as discussed previously, there are many exceptions to this general rule. For commodities sensitive to low-temperature injury, temperatures just above freezing will result in disorders and decay. Recommended storage conditions, therefore, become a compromise to obtain the maximum time at an economical cost without any adverse reaction to the imposed environmental conditions.

When consulting published tables of recommended storage conditions, it needs to be remembered that such tables have been compiled from data generated by many research groups in many countries. Therefore, they should only be used as a guide to likely storage behaviour for any particular commodity. The precise optimal storage conditions for a specific variety of a particular commodity grown in a specific locality need to be determined experimentally.

Temperature

Recommended general storage temperature ranges for a selection of fruits and vegetables are presented in Table 13.1. The storage period ranges listed under each of the three storage temperature ranges recognise that for a particular product, factors such as cultivar, season and maturity at harvest can give different physiological and metabolic responses, which result in variations in the optimum storage temperature. Moreover, the optimum storage temperature can be slightly outside the ranges specified in Table 13.1. For instance, optimum storage temperatures reported for apples range from –1°C to 5°C. The temperature groupings in Table 13.1 are useful when the volume of a specific produce line is insufficient to fill a storage room or transport container. A number of different types of produce may then need to be held in the same storage chamber, so a temperature compromise is required. Generally safe temperature settings for long-term mixed storage of horticultural produce could be 1°C for produce that is not susceptible to low-temperature injury, 5°C for produce that is susceptible to low-temperature disorders, and 10°C for produce that is susceptible to chilling injury. For produce that is highly susceptible to chilling injury or where further maturation is required, ambient storage is recommended (see Chapter 4). Considerations of compatibility of humidity requirements (see Chapter 5) and ethylene sensitivity (see Chapter 6) will also be important in determining the feasibility of mixed-storage arrangements.

It is important to bear in mind that a time by temperature relationship exists for the development of low-temperature injuries. Accordingly, low temperature–sensitive produce can often be safely held at sub-optimal temperatures for short periods (see Chapter 8). This means that where short-term mixed storage for up to about 1 week is required, such as in transit to markets or holding in wholesale or retail markets, a more liberal temperature regime can apply. Thus, a large number of products can be safely stored together at 0–1°C, with most others being held safely at 5–10°C. Some tropical produce, potatoes, onions and pumpkins may be best held at 15°C, or even at ambient temperature.

Table 13.1 Recommended temperature to maximise storage life of selected fruit and vegetables

	Time at optimum temperature (weeks)				Time at optimum temperature (weeks)		
	–1–4°C	5–9°C	10°C+		–1–4°C	5–9°C	10°C+
FRUIT				VEGETABLES			
Very perishable (0–4 weeks)				*Very perishable (0–4 weeks)*			
Apricot	2			Asparagus	2–4		
Banana, green			1–2	Bean	1–3		
Berry fruits	1–2			Broccoli	1–2		
Cherry	1–4			Cucumber		2–4	
Fig	2–3			Lettuce	1–3		
Mango			2–3	Mushroom	2–3		
Strawberry	1–5 days			Pea	1–3		
				Tomato			1–3
Watermelon		2–3		*Perishable (4–8 weeks)*			
Perishable (4–8 weeks)				Cabbage	4–8		
Avocado		3–5		*Semi-perishable (6–12 weeks)*			
Grape	4–6			Celery	6–10		
Mandarin		4–6		Leek	8–12		
Passionfruit		4–5		Marrow			6–10
Peach	2–6			*Non-perishable (>12 weeks)*			
Pineapple, green			4–5	Carrot	12–20		
Plum	2–7			Onion	12–28		
Semi-perishable (6–12 weeks)				Potato		16–24	
Coconut	8–12			Pumpkin			12–24
Orange		6–12		Sweet potato			16–24
Non-perishable (>12 weeks)							
Apple	8–30						
Grapefruit			12–16				
Pear	8–30						

The storage times presented for various products in Table 13.1 show a wide variation, ranging from a few days to many months. While the storage life of individual commodities is governed by many factors, an over-riding determinant as to whether they have a short or long storage life is the overall rate of metabolism, which is related to the respiration rate. Produce with a low respiration rate can generally be stored for longer. Some examples of produce respiration rates are given in Table 13.2. The produce with the shortest storage life are leafy vegetables with their high surface area to volume ratio, fruits that are harvested when ripe and rapidly metabolising (e.g. berries), and chilling-sensitive tropical fruit that cannot be held at low temperature to decrease its metabolic rate. Produce with the longest storage life are generally underground vegetables and those pome and citrus fruits with a relatively low respiration rate and/or which can tolerate low-temperature storage.

Table 13.2 Typical respiration rates of selected fruit and vegetables

Fruit		Respiration rate at 15°C (mL CO_2 kg^{-1} h^{-1})	Vegetable	Respiration rate at 15°C (mL CO_2 kg^{-1} h^{-1})
Apple		25	Bean	250
Banana	green	45	Cabbage	32
	ripe	200	Carrot	45
Grape		16	Lettuce	200
Orange		20	Pea	260
Peach		50	Potato	8
Pear		70		
Strawberry		75		

SOURCE Adapted from American Society of Heating, Refrigerating and Air-conditioning Engineers (1986) *ASHRAE handbook of refrigeration systems and applications*. ASHRAE: Atlanta GA, USA.

Termination of storage life in many products, even at low temperatures, is often associated with the onset of decay. The ability of products to withstand microbial invasion is related not only to external environmental conditions (e.g. humidity), but also to the structural integrity of cells comprising the host tissue. Thus, normal metabolism during storage that leads naturally to senescence will lead in turn to increased germination of spores on the host surface and/or to release from quiescence of existing latent infections (see Chapter 9).

Humidity

A high RH around produce is required to minimise water loss. However, as discussed in Chapter 5, the use of nearly 100% RH can create problems due to condensation of water. Free water promotes spore germination for many fungal pathogens and also stimulates bacterial growth, both of which may lead to rots. A saturated atmosphere can only be used where produce has some resistance to rotting and/or is prone to excessive rates of water loss. Recommended humidity conditions then become a compromise between reducing water loss and preventing microbial growth, with the latter factor being of paramount importance. For most produce, RH recommendations tend to be in the range of 85–95%, but they can be about 98% for produce with very high transpiration rates (e.g. leafy vegetables) and around 60% for produce highly susceptible to rotting (e.g. onions).

Commodity group recommendations

Leafy vegetables and immature flower heads

These commodities have high transpiration rates due to their large surface area to volume ratios. Moreover, many, particularly those comprising immature tissues, have high rates of respiration. They should be cooled to about 1°C as soon as practical after harvest and maintained at low temperature throughout storage, and preferably also during marketing. The RH should be at least 95%. It has been suggested that 100% RH can be sustained provided that the temperature is not higher than 1°C. Some consideration needs to be given to removing the heat of respiration from more immature vegetables, which may require ventilation with high-humidity air.

Vegetable fruits

Vegetable fruits can be divided into two sub-groups: those consumed in an immature unripe condition (e.g. green legumes, cucumber and peppers), and those consumed when mature and ripe (e.g. melons, pumpkin and tomato). Most vegetable fruits have some susceptibility to low-temperature injuries and are, therefore, stored at temperatures ranging from 3–5°C for beans to 10–15°C for pumpkins. There are exceptions, such as green peas, which can be stored at 0°C. Tomato, in common with many other climacteric fruits, is more tolerant of low temperatures when ripe. A high RH (greater than 95%) is essential for peas and beans, but a low RH (about 60%) is required for pumpkin and winter squash, with most other vegetable fruits being held at 90–95%.

Underground vegetables

This group encompasses diverse organ types, including bulbs (e.g. onion), roots (e.g. carrot), tubers (e.g. potato) and rhizomes (e.g. ginger). Most underground organs are characterised by a low respiration rate, and thus tend to have a relatively long storage life. Bulbs can be stored at 0°C or slightly lower, but need to have a low RH (about 70%) to prevent rotting and/or root growth. Bulbs should be cured before long-term storage (see Chapter 11). They should not be held for extended periods at 5–20°C, as rotting and sprouting occur rapidly at these temperatures. Most of the temperate root vegetables should be stored at 0°C and 95% or higher RH. Sweet potato, which is a fleshy storage root, is susceptible to chilling injury at temperatures below 10°C and benefits from curing before storage. Most tubers are susceptible to low-temperature injuries and should be held at temperatures from 5–15°C. Sprouting and rotting are the major storage problems and most tubers benefit from curing.

Deciduous tree and vine fruit

The storage life of deciduous fruit tree produce is highly variable, ranging from 1–2 weeks for apricot and figs to over 6 months for some apples and pears (Table 13.1). Temperate fruits with an inherently short storage life benefit from rapid cooling after harvest. The recommended storage temperature for most deciduous and vine fruit is –1 to 0°C. The exceptions are specific apple and pear cultivars that are susceptible to physiological disorders, where the recommended temperature can be up to 5°C. While deciduous fruits with a relatively short storage life are also generally susceptible to low-temperature injuries, 0°C remains the recommended temperature as an even shorter storage life, through enhanced senescence, occurs at slightly higher temperatures. Care must be taken with deciduous fruit stored in an unripe condition, as prolonged storage at low temperatures can inhibit the ability of fruit to ripen after cool storage. European pear cultivars are particularly susceptible to this effect. The RH during storage of deciduous and vine fruit should be 90–95%.

Berries

As a group, berries are probably the most perishable of all fruit. Some, such as blackberries, have only a few days of storage life even under optimal conditions. Berries are non-climacteric fruit with high respiration rates and so are harvested at optimum eating quality. Their soft and succulent nature renders them highly susceptible to physical damage that leads to general senescence and rotting. The recommended storage temperature for most berries is –1 to 0°C at 90–95% RH. Berries benefit from rapid cooling soon after harvest.

Citrus fruit

Citrus fruits are prone to a wide range of physiological disorders. Susceptibility to these disorders varies greatly among species, between cultivars and often across growing regions (see Chapter 8). The genotype-associated variability gives rise to a wide range of recommended storage temperatures, from 4°C for mandarin to 15°C for grapefruit. Rotting and water loss can be problems during prolonged storage, and RH of 85–90% is recommended. Waxes with added fungicides are often applied to overcome these problems (see Chapter 9).

Tropical and subtropical fruit

Most tropical and subtropical fruit is susceptible to low-temperature injuries, with the degree of severity often being related to temperature conditions prevailing in the production environment (see Chapter 8). Most tropical fruit is highly susceptible and should not be stored below 10°C. Some subtropical fruit, such as avocado, has a wide range of recommended temperatures, from 4 to 13°C. Kiwifruit can be stored at 0°C. Ripe fruit can often be stored at temperatures about 5°C lower than unripe fruit, albeit with a more limited storage life. The recommended humidity for tropical and subtropical fruits is in the range 85–95% RH.

Ornamentals

Fresh cut flowers and foliage often comprise immature organs, such as flower buds and young leaves. They generally have high surface area to volume ratios. Consequently, cut flowers and foliage have fairly high metabolic rates and high rates of water loss. Accordingly, they should be stored at the lowest temperature they can tolerate and at high RH. Generic storage recommendations for such produce are typically 1°C and 95% RH or higher. Such conditions are applicable to the major commercial cut flower crops (e.g. rose, carnation). However, a number of cut flower and foliage lines, particularly those of subtropical and tropical origin, are chilling-sensitive. Chilling-sensitive species should be held at temperatures of 13°C or above. Temperatures in the range 3–7°C are more appropriate for species, like kangaroo paw, that are moderately susceptible to low-temperature injury. The optimum temperature thus depends on the species, cultivar, stage of maturity, production environment, and duration of storage. Cut flowers and foliage can be stored wet (i.e. standing in solution) or dry. For longer-term storage, the dry option is preferred because it is more space efficient, metabolism of the plant may be lower, and the potential for decay is potentially reduced. Cut flowers may be given various treatments prior to storage, for example they may be pulsed with sucrose solution, to help maintain quality. Recommended storage conditions for a selection of cut flowers are given in Table 13.3.

Table 13.3 Storage conditions for selected cut flowers

Cut flower	Temp. (°C)	RH (%)	Storage life (days)	Short-term storage temp. (°C)
Alstroemeria*	0–4	90–95	6–10	1
Anthurium*	12.5–15.5	90–95	3–10	15
Bird-of-paradise	7–10	85–95	3–28	7.5
Carnation*	0–7	90–95	3–42	1
Chrysanthemum	–0.5–8	90–98	7–42	1
Delphinium*	0–4.5	90–95	1–2	1
Freesia*	0–4	90–95	1–14	1
Ginger	7–10	90–95	5	7.5
Gypsophila*	0–4.5	98	1–21	1
Iris§	–0.5–4	90–95	4–28	1
Liatris	0–5	90–95	3–14	1
Lily*	0–4.5	90–95	4–28	1
Lisianthus	1	90–95	7	1
Narcissus§	0–2	90–95	7–21	1
Orchid†	0.5–15	90–95	7–28	7.5
Protea	2–4	95	21	2
Rose*	0–4	90–98	4–14	1
Snapdragon*	–1–5	95	3–28	1
Statice	1.5–4	90–95	14–42	2
Tulip§	–0.5–2	85–95	3–42	1

* Cut flowers for which STS pulsing has been recommended.

§ Cut flowers reported to be sensitive to ethylene, but for which specific STS recommendations have not been made.

† Tropical orchids (e.g. vanda, dendrobium, cattleya) may be stored at 5–15°C, whereas cymbidiums and paphiopedilum may be stored at lower temperatures (e.g. –0.5–4°C).

SOURCES Principal reference: R.E. Hardenburg, A.E. Watada and C.Y. Wang (1990) *The commercial storage of fruits, vegetables and florist and nursery stocks.* (rev. ed.) US Department of Agriculture: Washington DC, USA. Handbook no. 66. Also: R. Jones (1991) *Post-harvest care of cut flowers.* Institute of Plant Sciences, Department of Agriculture: Knoxfield, Victoria; D. Joyce (1988) *Storage conditions for ornamental crops*, Western Australian Department of Agriculture: South Perth. Farmnote no. 34/88; J. Nowak and R.M. Rudnicki (1990) *Postharvest handling and storage of cut flowers, florist greens and potted plants*, Timber Press: Portland OR, USA.

Potted plants are generally not stored. However, if storage is necessary, a temperature of 13°C or higher is generally recommended. Since the range of pot plant species is broad, some require storage at relatively low temperatures of about 5°C, and some at high temperatures of about 20°C. Examples of recommended storage conditions for selected foliage lines and pot plants are given in Table 13.4, and for a number of bulbs in Table 13.5.

Table 13.4 Recommended storage temperature for selected cut foliage and pot plants

Plant	Temperature (°C)
*A. Cut foliage**	
Asparagus	0–5
Dieffenbachia	13
Eucalyptus	1.5–5
Holly	0–2
Maidenhair fern	0–4.5
B. Pot plants†	
Aglaonema	16–18
Begonia	10
Chrysanthemum	0–10
Maidenhair fern	15–18
Kalanchoe	10

* Suggested holding temperatures for cut foliage for at least short-term (<7 days) storage under high humidity (e.g. 90–95%+) conditions.

† RH in the range 65–90% has been recommended for pot plants.

Table 13.5 Recommended storage temperature and storage life for selected bulbs*

Plant		Temperature (°C)	Storage period (months)
Alstroemeria		4.5–10	–
Caladium		10–15.5	–
Freesia		22–30	3–4
Gladiolus		3.5–10	5–8
Iris		16–20	4–12
Lily, Easter		−0.5–0.5	10
Narcissus		7–20	2–4
Tulip	forcing	4.5–10	2–4
	outdoors	−0.5–0	5–6
Zantedeschia		2–4.5	–

* RH in the range 70–90% is suitable for most bulbs. Air movement (ventilation) around dry stored bulbs is desirable.

Many ornamental crops are very sensitive to ethylene (see Chapter 6), showing accelerated senescence or abscission. Ethylene-sensitive crops, such as carnations, can either be treated prior to storage with anti-ethylene compounds (e.g. silver thiosulphate) or stored in systems offering ethylene venting and/or scrubbing.

Atmosphere control

The postharvest life of horticultural produce held at optimum temperature and humidity conditions can be further extended by controlling the concentrations of carbon dioxide, oxygen and ethylene in the surrounding atmosphere. The optimum levels of carbon dioxide and oxygen are documented for many products (see Chapter 6), but particular storage situations and marketing logistics will determine whether or not it is commercially feasible to implement controlled atmosphere conditions. Recent studies have indicated that some produce may be sensitive to much lower concentrations of ethylene in the storage environment than previously believed. Thus, stringent removal of ethylene may be warranted.

Glossary of botanical names

Common and botanical names of some fruits and vegetables

Common name		Botanical name
Apple		*Malus x domestica* Borkh.
Apricot		*Prunus armeniaca* L.
Asian pear		*Pyrus pyrifolia* Nakai and *P. bretschneideri* Rehder
Asparagus		*Asparagus officinalis* L.
Avocado		*Persea americana* Mill.
Banana		*Musa* L. sp. Cavendish varieties *M. acuminata* Colla
Beans	broad	*Vicia faba* L.
	string	*Phaseolus vulgaris* L.
	mung	*Phaseolus aureus* Roxb.
Beetroot		*Beta vulgaris* L.
Blueberry		*Vaccinium* sp.
Broccoli		*Brassica oleracea* L. (Italica group)
Brussels sprout		*Brassica oleracea* L. (Gemmifera group)
Cabbage		*Brassica oleracea* L. (Capitata group)
Carambola		*Averrhoa carambola* L.
Carrot		*Daucus carota* L.
Cassava (manioc, tapioca)		*Mannihot esculenta* Crantz
Cauliflower		*Brassica oleracea* L. (Botrytis group)
Celery		*Apium graveolens* L.
Cherimoya		*Annona cherimola* Mill.
Cherry	sweet	*Prunus avium* L.
	sour	*Prunus cerasus* L.

Chilli	*Capsicum annuum* L.
Choko	*Sechium edule* (Jacq.) Sw.
Corn (maize), sweet	*Zea mays* L.
Cucumber	*Cucumis sativus* L.
Eggplant (aubergine)	*Solanum melongena* L.
Feijoa	*Feijoa sellowiana* Berg.
Fig	*Ficus carica* L.
Garlic	*Allium sativum* L.
Ginger	*Zingiber officinale* Rascoe
Globe artichoke	*Cynara scolymus* L.
Grape	*Vitis vinifera* L.
Grapefruit	*Citrus paradisi* Macfad.
Guava	*Psidium guajava* L.
Jackfruit	*Artocarpus heterophyllus* (Lam.) L.
Jerusalem artichoke	*Helianthus tuberosus* L.
Kiwifruit	*Actinidia deliciosa* (A. Chev.) C.F. Liang & A.R. Ferguson
Leek	*Allium ampeloprasum* L.
Lemon	*Citrus limon* (L.) Burm. f.
Lettuce	*Lactuca sativa* L.
Lime	*Citrus aurantifolia* (Christm.) Swingle
Litchi	*Litchi chinensis* Sonn.
Loquat	*Eriobotrya japonica* Lindl.
Mandarin	*Citrus reticulata* Blanco
Mango	*Mangifera indica* L.
Mangosteen	*Garcinia mangostana* L.
Muskmelon (cantaloupe, honeydew)	*Cucumis melo* L.
Nectarine	*Prunus persica* (L.) Batsch.
Okra	*Hibiscus esculentus* L.
Onion	*Allium cepa* L.
Orange, sweet	*Citrus sinensis* (L.) Osbeck
Papaya	*Carica papaya* L.
Parsley	*Petroselinum crispum* (Mill.) Nym.
Parsnip	*Pastinaca sativa* L.
Passionfruit	*Passiflora edulis* Sims
Pea	*Pisum sativum* L.
Peach	*Prunus persica* (L.) Batsch.
Pear	*Pyrus communis* L.
Pepino	*Solanum muricatum* Ait.
Peppers, green and red	*Capsicum annuum* L.
Persimmon	*Diospyros kaki* L.f.
Pineapple	*Ananas comosus* (L.) Merr.
Plum	*Prunus domestica* L.
Pomegranate	*Punica granatum* L.
Potato	*Solanum tuberosum* L.
Pumpkin	*Cucurbita pepo* L.
Radish	*Raphanus sativa* L.

Rambutan	*Nephelium lappaceum* L. var. *esculentum* Nees
Rhubarb	*Rheum* sp.
Satsuma mandarin	*Citrus unshu* Mari
Soya bean	*Glycine max* (L.) Merr.
Spinach, European	*Spinacia oleracea* L.
Squash	*Cucurbita maxima* Duch.
Strawberry	*Fragaria* x *ananassa* Duch.
Swede turnip	*Brassica napus* L. (Napobrassica group)
Sweet potato	*Ipomea batatas* (L.) Lam.
Tamarillo (tree tomato)	*Cyphomandra betacea* (Cav.) Sendt.
Taro	*Colocasia esculenta* (L.) Schott
Tomato	*Lycopersicon esculentum* Mill.
Turnip	*Brassica campestris* L. (Rapifera group)
Watermelon	*Citrullus lanatus* (Thunb.) Mansf.
Yam	*Dioscorea batatas* Deene

Common and botanical names of some cut flowers and foliage

Common name	Botanical name
Alstroemeria, lily of the Incas, Peruvian lily	*Alstroemeria* L. spp.
Anthurium, tailflower	*Anthurium andreanum* Linden ex Andre
Belladonna lily	*Amaryllis belladonna* L.
Bird-of-paradise, strelitzia	*Strelitzia reginae* Banks & Dryander
Carnation	*Dianthus caryophyllus* L.
Cattleya orchid	*Cattleya* Lindley spp.
Christmas mistletoe	*Phoradendron tomentosum* (DC.) Engelm. ex A. Gray
Chrysanthemum, florist's chrysanthemum	*Dendranthema* (DC.) Des Moul. spp.
Cymbidium orchid	*Cymbidium* Sw. spp.
Delphinium, larkspur	*Delphinium* L. hybrids
Dendrobium orchid	*Dendrobium* Sw. spp.
Eucalyptus, gum tree	*Eucalyptus* L'Herit. spp.
Freesia	*Freesia* x *hybrida* L. Bailey
Gerbera, Transvaal daisy	*Gerbera jamesonii* Bolus ex Hook f.
Ginger	*Alpinia* Roxb. spp.
Grevillea	*Grevillea* R. Br. ex J. Knight spp.
Gypsophila, baby's breath, gyp	*Gypsophila paniculata* L.
Iris, bulbous iris	*Iris* L. hybrids, Dutch
Kangaroo paw	*Anigozanthos* Labill. spp.
Liatris, gayfeather	*Liatris spicata* (L.) Willd.
Lily	*Lilium* L. spp.
Limonium, statice	*Limonium* Miller spp.

Lisianthus, Texas rose	*Eustoma grandiflorum* (Raf.) Shinn.
Narcissus, daffodil	*Narcissus* L. hybrids
Paper daisy	*Helipterum* DC. spp.
Paphiopedilum orchid	*Paphiopedilum* Pfitzer spp.
Protea	*Protea* L. spp.
Rose	*Rosa* L. hybrids
Snapdragon	*Antirrhinum majus* L.
Tulip	*Tulipa* L. hybrids
Vanda orchid	*Vanda* Jones ex R. Br. spp.

Common and botanical names of some pot plants

Common name	Botanical name
African violet, Saintpaulia	*Saintpaulia ionantha* H. Wendl.
Azalea, Indian azalea	*Rhododendron* L. spp.
Begonia	*Begonia* L. hybrids
Boston fern	*Nephrolepis exaltata* (L.) Schott
Chrysanthemum	*Dendranthema* (DC.) Des Moul spp.
Dracaena	*Dracaena* Vand. ex L. spp.
Ficus, fig, rubber tree	*Ficus* L. spp.
Frangipani, plumeria	*Plumeria* L. spp.
Hibiscus	*Hibiscus* L. spp.
Ivy	*Hedera* L. spp.
Kalanchoe, Palm Beach bells	*Kalanchoe* Adams spp.
Melaleuca	*Melaleuca* L. spp.
Poinsettia	*Euphorbia pulcherrima* Willd. ex Klotzsch
Schefflera	*Schefflera* S. Forster & F. Forster
Spathiphyllum, spathe flower	*Spathiphyllum* Schott spp.
Yucca	*Yucca* L. spp.

Further reading

Books

Abeles, F.B., P.W. Morgan and M.E. Saltveit (1992) *Ethylene in plant biology* (2nd ed.), Academic Press: New York NY, USA.

Bartz, J.A. and J.K. Brecht (eds) (2003) *Postharvest physiology and pathology of vegetables* (2nd ed.), Marcel Dekker: New York NY, USA.

Buchanan, B.B., W. Gruissem and R.L. Jones (eds) (2006) *Biochemistry and molecular biology of plants,* American Society of Plant Biologists. John Wiley: West Sussex, UK.

Burg, S.P. (2004) *Postharvest physiology and hypobaric storage of fresh produce,* CABI: Cambridge MA, USA.

Kader, A.A. (ed.) (2002) *Postharvest technology of horticultural crops* (3rd ed.), Publication 3311. University of California, Division of Agriculture and Natural Resources: Oakland CA, USA.

Kays, S.J. and R.E. Paull (2004) *Postharvest biology,* Exon: Athens GA, USA.

Lawless, H.T. and H. Heyman (1999) *Sensory evaluation of food: Principles and practices.* Aspen Publishers: Frederick MD, USA.

Little, C.R. and R.J. Holmes (2000) *Storage technology for apples and pears: a guide to production, postharvest treatment and storage of pome fruit in Australia* (ed. Faragher, J.) Department of Natural resources and Environment: Victoria, Australia.

Seymour, G.B., J.E. Taylor and G.A. Tucker (eds) (1993) *Biochemistry of fruit ripening,* Chapman & Hall: London, UK.

Shewfelt, R.L. and S.E. Prussia (1993) *Postharvest handling: A systems approach,* Academic Press: San Diego CA, USA.

Snowdon, A.L. (1990) *A colour atlas of post-harvest diseases and disorders of fruits and vegetables.* Vol 1, *General introduction and fruits,* Wolfe Scientific: London, UK.

Snowdon, A.L. (1991) *A colour atlas of post-harvest diseases and disorders of fruits and vegetables.* Vol 2, *Vegetables,* Wolfe Scientific: London, UK.
Thompson, A.K. (2003) *Fruit and vegetables. Harvesting, handling and storage,* (2nd ed.) Culinary and Hospitality Publications Services: Weimar TX, USA.
Watkins, J.B. (1990) *Forced-air cooling* (2nd ed.). Information series UQ188027, Queensland Department of Primary Industries: Brisbane, Australia.

Websites

Food composition tables: www.fao.org/infoods/directory_en.stm
Gross, K. and H.A. Wallace, *The commercial storage of fruits, vegetables and florist and nursery stocks*: http://usna.usda.gov/hb66
Postharvest Technology Research and Information Center, University of California, Davis CA, USA: http://postharvest.ucdavis.edu
Quality assurance systems. BRC (British Retail Consortium): www.brc.org.uk; EurepGAP: www.eurep.org; Freshcare Ltd: www.freshcare.com.au; GFSI (The Global Food Safety Initiative): www.ciesnet.com; SQF Institute: www.sqfi.com
Stewart Postharvest Review: www.stewartpostharvest.com
Sydney Postharvest Laboratory: www.postharvest.com.au/Default.html
Watkins, C.B. and W.M. Miller, A summary of physiological processes or disorders in fruits, vegetables and ornamental products that are delayed or decreased, increased or unaffected by application of 1-methylcyclopropene (1-MCP): http://www.hort.cornell.edu/mcp/ethylene.pdf

Index